DISCRETE OR CONTINUOUS?
The Quest for Fundamental Length in Modern Physics

The idea of infinity plays a crucial role in our understanding of the universe, with the infinite spacetime continuum perhaps the best-known example – but is spacetime *really* continuous? Throughout the history of science, many have felt that the continuum model is an unphysical idealization, and that spacetime should be thought of as "quantized" at the smallest of scales.

Combining novel conceptual analysis, a fresh historical perspective, and concrete physical examples, this unique book tells the story of the search for the fundamental unit of length in modern physics, from early classical electrodynamics to current approaches to quantum gravity. Original philosophical theses, with direct implications for theoretical physics research, are presented and defended in an accessible format that avoids complex mathematics. Blending history, philosophy, and theoretical physics, this refreshing outlook on the nature of spacetime sheds light on one of the most thought-provoking topics in modern physics.

AMIT HAGAR is an Associate Professor in the Department of History and Philosophy of Science, Indiana University Bloomington, where he specializes in the foundations of modern physics.

DISCRETE OR CONTINUOUS?

The Quest for Fundamental Length in Modern Physics

AMIT HAGAR

Indiana University

CAMBRIDGE
UNIVERSITY PRESS

CAMBRIDGE
UNIVERSITY PRESS

University Printing House, Cambridge CB2 8BS, United Kingdom

Cambridge University Press is part of the University of Cambridge.

It furthers the University's mission by disseminating knowledge in the pursuit of education, learning and research at the highest international levels of excellence.

www.cambridge.org
Information on this title: www.cambridge.org/9781107633698

First published 2014
First paperback edition 2015

A catalogue record for this publication is available from the British Library

ISBN 978-1-107-06280-1 Hardback
ISBN 978-1-107-63369-8 Paperback

To my parents

Time is nature's way of keeping everything from happening at once;
space is what prevents everything from happening to me.

Attributed to John Wheeler

Contents

Preface

How does one end up writing a book on length, you may ask? Truth be told, several years ago I took issue with the quantum information community on matters foundational and practical, such as the role of decoherence in the construction of a large scale and fault tolerant quantum computer [Hag10], and in so doing realized how much physics could benefit from the history and the philosophy of science. Philosophical problems of spacetime physics have always been at the back of my mind, ever since I wrote my MA thesis on the Hole Argument (with the late Itamar Pitowsky) and my PhD thesis on Time's Arrow (under the supervision of Steven Savitt), and I could see that a similar lesson could be learnt in the domain of quantum gravity.

And so began this project on fundamental length. Its seeds were sown in the same place most of my previous ideas for research projects have come from for the last five years or so, namely, in Jeffery Bub's and Robin Schuster's backyard, after a long day of listening to talks by some of the brightest minds in the field of the foundations of physics at the annual "New Directions" workshop, unwinding with a bottle of cold beer and enjoying Jeff's and Robin's warm hospitality on one of DC's crisp spring evenings. Earlier that morning a session had been dedicated to quantum field theories, and the issue of a cutoff (a bound on momentum transfer) kept emerging as crucial to the interpretation of the theory. As I would find out later, and as you will if you read down these pages, this cutoff would become one of the main themes of this book.

That summer, stuck in a remote and secluded Alpine hut, high above Valsugana, not far from Trento, I delved into the history of the continuous versus the discrete, and discovered a whole new world. It was as if millennia-old debates and arguments from the philosophy of mathematics were reincarnated in discussions on the role of fundamental length in field theories, and moreover, as if these discussions from the 1930s were reappearing today in the attempts to construct a quantum theory of gravity. I wrote a concise grant proposal, delineating the benefits of unraveling this *terra incognita*, and went down from the mountain back to the madding crowd.

Many have contributed to the project in the four years that followed. First and foremost, Fred Kronz from the NSF, which generously supported the project from its outset, under grant # SES 0951179 (at this point I should also mention that any opinions, conclusions or recommendations expressed in this book are mine alone, and do not necessarily reflect the views of the NSF). Next, my colleagues from Indiana University, especially Sandy Gliboff (HPS), who helped me with the translation from German of Heisenberg's and Schrödinger's less known papers, as well as Wataghin's papers, The IU Center for Spacetime Symmetries and in particular Alan Kostelcky (physics), who opened a door to the Standard Model Extension and the tests for CPT violations, Gerardo Oritz (physics) who patiently explained to me how things appear from the condensed matter perspective, and Yuri Bonder (physics).

I am also grateful to C. Alden Mead, now retired from UMN, for revealing, almost 50 years after their publication, the story behind his papers on fundamental length; to Bruno Carazza, now retired from the University of Parma, for sharing his experience as a student of Gleb Wataghin, and for his interest and support of the project; to Roberto Giuntini and his talented group at the University of Cagliari, where the first two chapters of this book were written under the hospitality of the Reggione della Sardegna Visiting Professor Program; to Bill Unruh (UBC, physics), whose keen eye for physical fallacies is the sharpest I have ever known; to Tom Banks (Rutgers, physics) who patiently explained to me the intricacies of black hole thermodynamics; to Sabine Hossenfelder, a quantum gravity phenomenologist from Nordita, who runs a popular blog at backreaction.blogspot.com; to Giovanni Amelino-Camelia (La Sapienza, physics); to Olival Freire Jr. and Saulo Carneiro from the Institute of Physics at the Universidade Federal da Bahia, where the final chapter of the book was written, for their hospitality and for discussions; to Osvaldo Pessoa (USP, philosophy) for arranging a meeting with the late Amelia Hamburger, and for getting me access to Wataghin's archives at USP; to Bill Demopoulos (UWO, Canada) and Meir Hemmo (U Haifa, Israel), for being my mentors and friends for so many years, and to the anonymous referees for their comments and their criticism that helped me sharpen my ideas.

Parts of the material in this book – the Einstein–Swann correspondence, the critique on the argument for the consistency of loop quantum gravity with the Lorentz contraction, and the critique on the "disappearance of spacetime" in theories of quantum gravity – were published in slightly different versions in *Studies in the History and the Philosophy of Modern Physics* [Hag08, Hag09, HH13]. The work presented here is naturally more comprehensive, placing as it does these separate issues in their common context.

Finally, this book could probably have been finished sooner than it actually was without my wife *Adele* and my daughter *Aviv*, but then, it wouldn't have been worth it, would it?

1

Introduction

A famous anecdote about Gauss would have him measuring the angles of the great triangle formed by the mountain peaks of Inselberg, Brocken, and Hohenhagen, while looking for evidence that the geometry of space is non-Euclidean. Whether true or simply a myth, this oft told story emphasizes the tug of war that has existed between Plato, Kant, Poincaré and other idealists, *a priorists* and conventionalists, and those who have insisted that the geometrical structure of our world is an empirical matter of fact, to be decided by experience.

For many years the debate was confined to the question of curvature. Non-Euclidean geometries were initially believed to be impossible, and even after their discovery, they were still considered inappropriate for describing the geometry of three-dimensional space. Flat (Euclidean) geometry, on the other hand, was always the common view, chosen by default as ideal, or *a priori*, or simply more convenient.

The history of science has taught us otherwise. As Einstein once said [Ein29], "People slowly accustomed themselves to the idea that the physical states of space itself were the final physical reality." The geometry of this reality turned out to be constrained by the distribution of matter, hence by empirical, contingent, matters of fact. Moreover, the general theory of relativity that describes it has turned physical objects (matter fields) into geometrical objects, and geometrical objects (the metric and the affine connection) into physical objects.

In this book we shall focus on another feature of geometry that exemplifies this tug of war between the *a priorists* and the empiricists. Rather than curvature, we shall be dealing with the structure of the line segment to which one-dimensional space is believed to be isomorphic. To the naked eye this one-dimensional space looks continuous, much as a film made of 24 frames per second does, but is it truly so at its most fundamental level?

Of course, there is no way to answer this question by direct observation. As humans, we are bounded by finite resolution capabilities; no matter how fine grained our "rulers" may be, their resolution can never directly reveal the putative fundamental continuum of the physical line segment. This line segment may be fundamentally discrete (and may appear discrete to our most refined "rulers"), or it may be fundamentally continuous (but still appear discrete to our most refined "rulers"). And yet, without evidence to the contrary, and consistent with our coarse grained (naked eye) observations, we indirectly *infer* a putative, continuous structure from the successful applicability of continuum geometry, the calculus, and the field concept with its continuous symmetries, to physical phenomena *in* space. This extrapolation is based on three premises: the mathematical structure with which we describe phenomena in physical space is continuous; the theories relying on this structure have been verified down to a scale of 10^{-16} cm; and we strongly believe that physical space is isomorphic to this structure, regardless of our actual finite resolution capabilities. Such an inference leaves plenty of room for debate on the possibility of alternative structures.

Our journey through the landscape of these alternative structures begins in Chapter 2, with the history and the philosophy of mathematics, where ingenious arguments have been constructed for (and against) spatial discreteness, as well as for (and against) its rival, the continuum. These arguments aim to establish that only one of these notions is logically consistent as a basis for geometry, and moreover, that only one of them could represent faithfully the structure of physical space.

The scope of these arguments is overwhelmingly vast. They range from topology to geometry, and from number theory to the theory of computation. But they all fail to establish their conclusion; there is nothing inconsistent in a geometry based on a discrete line segment (or on a continuous line segment, for that matter), and there is more than one way to describe the physical world. Notwithstanding the ubiquity of the calculus, spacetime may be discrete at the most fundamental level, and only appear continuous (that is, be faithfully represented by continuous geometry) at lower resolutions. Even more important, these arguments demonstrate the disagreement that prevails already with regard to the question of whether the character of the line segment that faithfully represents the physical world can be decided empirically. Throughout this book we shall clarify what it takes to answer this question.

More *a priori* arguments against finitism – treated here as a general philosophical thesis – are presented (and dismantled) in Chapter 3. These arguments target the putative philosophical price that finitism may carry, saddling it with metaphysical and epistemological commitments that appear to have dire consequences. For example, finitism may blur the distinction between metaphysics and epistemology by subordinating what *is* to what we *know*. All by itself this may not seem a serious allegation, but in the context of interpreting, say, the notion of probability in quantum mechanics, such blurring makes it difficult to distinguish quantum from classical probabilities.

Difficult, but not impossible: one can still identify the difference between classical and quantum probabilities in the structure of the respective probability spaces, i.e., whether this structure is Boolean or non-Boolean. This distinction sees quantum interference not as evidence for some metaphysical difference between classical and quantum "realities," but rather as representing, under several plausible physical assumptions, a quantitative difference in measure between the two. Remarkably, such a finitist and objective interpretation of quantum probability as a limit on measurement resolution is also consistent with the way the notion of fundamental length was introduced into modern physics, and it can be used to restore credibility to the thought experiments that motivated it, and to their putative implications.

Another angle on this tension between ontology and epistemology is that finitism may spill the baby with the tub water, as it threatens to turn the question of spatial discreteness into a metaphysical, undecidable question. If what *is* is dictated by what we *know*, then, or so the story goes, given our finite resolution capabilities, we could never decide whether the world is truly discrete or continuous. But here also, that the question of spatial discreteness is undecidable does not follow from the operational methodology that regards measurement results (which are always finite) as primitive. To see this, an operational argument that goes back to Einstein about the primacy of geometrical notions is shown to be distinct from the views of his contemporaries on the conventionality or the *a priori* character of geometry. We will return to this argument when discussing current approaches to quantum gravity.

Remote as these two chapters may seem from theoretical physics, they serve a double purpose. First, they bring to light a certain argument structure that repeats itself throughout the history of science, and emerges again and again in debates among physicists on the possibility of introducing fundamental length into field theories, and on the consistency of this notion with other well-established physical principles. Second, they emphasize the important question of whether or not the issue of spatial discreteness can be decided empirically. This question is at the heart of the ongoing debate on the quest for quantum gravity phenomenology.

Turning to the history of physics, we begin in Chapter 4 to follow the bumpy road along which the notion of fundamental length entered into field theories. Starting with classical electrodynamics and the problematic idea of the extended electron, via the divergence problems in quantum electrodynamics, and the remedies sought in momentum cutoffs and finite measurement resolution, we learn about the different roles the notion of fundamental length played for different physicists. Moving to quantum field theories we compare the finitist approach (that abhors actual infinity and singularity) with the renormalization program (and its acceptance of potential infinity). Along the way we encounter for the first time the main conceptual problems that accompany fundamental length – postulated at that time to be of the order of the electron radius – namely, non-locality, loss of relativistic causality, and tension with unitarity and Lorentz invariance. These problems (and the strategies for their possible solutions) resurface later in the attempts to "quantize" the general theory of relativity.

Here we also encounter for the first time the different ways in which the fundamental length was interpreted. The high momentum cutoff was initially seen as a limit on spatial resolution, and so the first, natural, interpretation was to regard it as a limit on the domain of applicability of the theory it was introduced to: the theory could not predict anything above the bound on momentum space (or below the bound in position space). Given the numerous problems that one encounters when trying to interpret the high momentum cutoff as signifying an actual spatial discreteness, and the fact that without additional premises there was no logical necessity for doing so, many physicists chose to remain agnostic with respect to the fundamental structure of space. Spatial discreteness thus remained an unsupported conjecture, even after the introduction of a fundamental length into quantum electrodynamics.

As is well known, partly because of the unsolved problems it created in field theories, and partly because of the remarkable experimental and theoretical success of the renormalization program, the notion of fundamental length remained at the margins of theoretical physics. And yet, that quantum field theory is not the arena in which the question of spatial discreteness can be decided follows not from this historical contingency, but from another simple fact, namely that quantum field theory as we know it does not cover all of physics; it leaves out gravitation. We therefore focus our attention on the ongoing attempts to write down a consistent theory of quantum gravity.

In Chapter 5 we lay out – to my knowledge, for the first time – the prehistory of these attempts. Motivated by relationalism and operationalism, mathematicians and physicists, some unknown to contemporary scholars, have entertained ideas germane to the solution of the quantum gravity problem – the problem of constructing a theory that could predict in a consistent way phenomena in domains where both gravitational and quantum field theoretic effects are believed to exist – and did so already in the late 1920s. These ideas include notions such as absolute uncertainty, non-commutative geometry, and non-trivial geometry of momentum space. In this chapter we also discuss the debate that emerged around the late 1950s on the necessity of quantizing the gravitational field, and its putative relation to the notion of fundamental length.

As a preamble to the discussion on current approaches to the quantum gravity problem, we take a short detour in Chapter 6 where we present and analyze a rarely mentioned correspondence between Einstein and the Anglo-American physicist W. F. G. Swann. In this correspondence Einstein reveals his insight into the limit of the "constructive," or "dynamical" approach to spacetime physics, which seeks to reduce spacetime and its geometrical features to relations between more fundamental building blocks and the dynamics thereof. In such an approach, one still needs to designate some degrees of freedom to represent metrical notions such as "length" or "duration," if only to make contact with experiments that can verify the said dynamics in spacetime. Therefore a strong reading of this approach, popular among quantum gravity theorists, according to which "spacetime disappears" at the fundamental level, is untenable. What is tenable is a weaker version of the dynamical approach, which sees the reduction process as a consistency proof. By this I mean a proof for the possibility that the dynamics of some postulated underlying building blocks are consistent with the observable spacetime structure, and may even help decide certain characteristic features of this structure, such as its symmetry groups or its metric. I call this insight "thesis L," and I demonstrate its applicability to Swann's unsuccessful attempts to "reduce" (in the strong sense above) the geometrical structure of Minkowski spacetime to the dynamics of quantum field theory.

We further demonstrate the validity of thesis L in Chapter 7, where we show that each of the current contenders for the solution of the quantum gravity problem succumbs to it by designating some degrees of freedom as representing a fundamental notion of "length," "area," or "volume," and does so by stipulation. In addition, each of these approaches either predicts or assumes that the structure of this fundamental geometrical notion is discrete, designating the fundamental length as the Planck length. We further inquire about the role this discreteness plays in these approaches, and find that very little has changed, at least from the conceptual perspective, since the 1930s. First, discreteness is still believed to be the cure for the divergences of quantum field theories, and now, since it includes gravity and hence is imposed almost 20 orders of magnitude below the scale of current observed interactions, it is also believed to be the cure for the singularities of the general theory of relativity. Second, the same difficulties that bothered the physicists who entertained the notion of fundamental length in the 1930s, for example, difficulties with locality, with relativistic causality, with unitarity, and with Lorentz invariance, still haunt us today.

Chapter 8 expands on this last point, focusing on two methodological challenges that quantum gravity theorists face today in their struggle to construct phenomenological models that could turn the question of spatial discreteness into an empirical question. First, it is extremely hard to sacrifice certain well-established principles that support the applicability of the continuous worldview at one – yet to be probed – scale (thus allowing the discrete model to yield new predictions) without violating them at all scales (thus making the discrete model false already). Second, settling for violations *in principle* of these well-established principles, which, given their negligibility or rarity, may never be tested *in practice*, is of no avail if one wishes to turn the question of spatial discreteness into a truly empirical question. In the final part of this chapter we find out whether or not these challenges are currently met.

A concise Q&A summary of the main theses presented and defended in the book can be found in Chapter 9.

Before we start, a disclaimer is in place. Throughout the history of science, many have entertained the notion of fundamental length and interpreted it either ontologically as physical spatial discreteness, or epistemologically as a limit on spatial resolution and on the domain of applicability of the field theory at hand. Owing to restrictions on length (excuse the pun), I have probably left out quite many of these (issues of propriety, for example, were not on my agenda). The historical account I present here mentions most of the major players and a few of the underdogs, yet is by no means exhaustive. Its sole purpose is to harness the history of physics for the sake of extracting philosophical and methodological morals, morals that may be useful to current research in quantum gravity. For "Those who cannot remember the past are condemned to repeat it" ([San05], p. 284).

2

Arguments from math

2.1 Outline

The basic mathematical tool, commonly used in the description of physical reality, has long been, and still is, the calculus. This tool is fundamentally continuous: it employs variables that range over continuous sets of values, and the functions it deals with are continuous. One such continuous function describes motion in space, and is treated as a function of a continuous time variable. The continuity of the motion function is essential, for velocity is regarded as the first derivative of this function, and acceleration as the second derivative. Functions which are not continuous are not differentiable, and hence they do not even have derivatives.

Despite its geometrical origins (e.g., [Boy49]), the calculus has been completely "arithmetized," and its development does not require any geometrical concepts. Nevertheless, as a tool, it is still applied to phenomena that occur in physical space. The applicability of the calculus to spatial events is achieved through analytic geometry, which begins with a one-to-one mapping between the points on a line and the set of real numbers. The set of real numbers constitutes a continuum in the strict mathematical (Cantorian) sense; consequently, the order-preserving one-to-one mapping between the real numbers and the points of the geometrical line renders the line a continuum as well. If, moreover, the geometrical line is a correct representation of lines in physical space, then physical space is likewise continuous.

The continuity of the calculus is thus buried deep in standard mathematical physics. Electromagnetism provides a typical example. Here, space is a structure S, diffeomorphic to $\mathbb{R}^3$, and the electromagnetic field at each point is an element of $Q = \mathbb{R}^6$, so that the phase space for the whole field is $Q^S = (\mathbb{R}^6)^{\mathbb{R}^3}$. On this phase space, we assign a dynamics in the form of a group of transformations T (time) indexed by $\mathbb{R}$. An impressive infinity for nature to keep track of.

While this is what the common view of physical space amounts to (see, e.g., Salmon in [Sal80], pp. 62–63), many throughout the history of science have felt that the continuum model is an unphysical idealization, and that there are in fact only a finite number of degrees of freedom in any finite volume:

It always bothers me that, according to the laws as we understand them today, it takes a computing machine an infinite number of logical operations to figure out what goes on in no matter how tiny a region of space and no matter how tiny a region of time. How can all that be going on in that tiny space? Why should it take an infinite amount of logic to figure out what one tiny piece of spacetime is going to do?[1]

([Fey65], p. 57)

This tension between the discrete and the continuous has its roots in the philosophy of mathematics, and goes back to the days of Zeno. Indeed, independently of its applicability to physical phenomena, the continuum is one of the most abstract concepts mathematicians have played with, and debates over its cogency have fueled the development of mathematical thought:

It will always remain a remarkable phenomenon in the history of philosophy, that there was a time, when even mathematicians, who at the same time were philosophers, began to doubt, not of the accuracy of their geometrical propositions so far as they concerned space, but of their objective validity and the applicability of this concept itself, and of all its corollaries, to nature. They showed much concern whether a line in nature might not consist of physical points, and consequently that true space in the object might consist of simple [discrete] parts, while the space which the geometer has in his mind [being continuous] cannot be such.

([Kan02], p. 288)

In this chapter we shall take a quick look at these debates, analyzing one of Zeno's arguments against the continuum, as well as other arguments from topology, geometry, and the theory of computation. These *a priori* arguments aim to establish – with varying degrees of sophistication – that only one mathematical structure – be it discrete or continuous – is logically possible, and, furthermore, that only one mathematical structure can be consistently applied as a description of the physical world.

Our goal here, however, will be to defend exactly the opposite claim. We shall try to demonstrate that (1) alternative mathematical descriptions of the line segment are as consistent as the standard, continuous one, and that (2) the question of the applicability of these structures to the physical world cannot be decided by logic alone.

---

[1] Feynman was introduced to the hypothesis of finite nature in general (and to the notion of cellular automata as a possible simulator to physics in particular) by Edward Fredkin around 1962 (Fredkin, private communication).

2.2 Zeno's paradox of extension

2.2.1 The paradox

One of the first people to have inquired about the nature of space (and time) from an *a priori* standpoint was a Greek named Zeno of Elea. A follower of Parmenides, Zeno sought to defend the metaphysical view that denied the possibility of change, motion, and plurality. Consequently, his paradoxes are viewed (e.g., in [Owe01]) as denouncing space, time, and motion as unreal and illusory. Together they form a double attack on *both* the continuous and the discrete.

Zeno's less famous paradox of plurality – also known as the paradox of extension – targets the notion of the continuum. It calls into question the consistency of this notion, and therefore it also underlies the four more famous paradoxes of motion. The latter raise further questions about the nature of motion as a functional relation between the two continua, space and time: Achilles and the Tortoise and the Dichotomy paradoxes target the view that motion is continuous; the Arrow and the Stadium target the view that motion is discrete.

We learn of the paradox of extension a full millennium later than Zeno, from what is supposed to be a direct quotation from him by Simplicus (sixth century AD):

In his [Zeno's] book, in which many arguments are put forward, he shows in each that a man who says that there is a plurality is stating something self-contradictory. One of these arguments is that in which he shows that, if there is a plurality, things are both large and small, so large as to be infinite in magnitude, so small as to have no magnitude at all. And in this argument he shows that what has neither magnitude nor thickness nor mass does not exist at all. For, he argues, if it were added to something else, it would not increase in size; for a null magnitude is incapable, when added, of yielding an increase in magnitude. And thus it follows that what was added was nothing . . .

The infinity of magnitude he showed previously by the same process of reasoning. For, having first shown that "if what is had not magnitude, it would not exist at all," he proceeds: "But, if it is, then each one must necessarily have some magnitude and thickness and must be at a certain distance from another. And the same reasoning holds good for the one beyond: for it also will have magnitude and there will be a successor to it. It is the same to say it once and to say it always: for no such part will be the last or out of relation to another. So if there is a plurality, they must be both large and small. So small as to have no magnitude, so large as to be infinite."

([Lee36], pp. 19, 21)

Zeno's paradox of extension presents the following dilemma: if plurality exists, i.e., if large things are composed of small ones, and infinite divisibility is allowed, then a finitely extended object is supposed to be an (infinite) sum of more basic parts. These basic parts can be extended or unextended. If these parts are unextended, then they have no magnitude, but then their (infinite) sum is also unextended. If these basic parts are extended, then their (infinite) sum is infinite. And so a finite, extended, object is "... both large and small. So small as to have no magnitude, so large as to be infinite."

The issue at stake here is, thus, the convergence (or lack thereof) of an infinite series. It is true that an infinite series of terms may converge to a finite sum when there is no smallest term in the series, but in Zeno's case the ultimate parts are all equal to some n where n is either 0 or some positive number. If the former is the case, the series does converge, but it converges to 0. If the latter is the case, the series diverges and has no finite sum.

2.2.2 Infinite divisibility

Since the paradox of extension relies on the notion of actual infinite divisibility, one way to resolve it is to deny this notion. Such a denial, however, must overcome the attempts to demonstrate mathematically the logical consistency of this notion that were well known to philosophers already in the seventeenth and eighteenth centuries.[2] These demonstrations fall into two categories: *reductio ad absurdum* proofs (where finite divisibility is assumed and an absurdity is supposed to follow), and proofs by construction. A famous example of the first type of proof is the (in)commensurability of the diagonal of a square with its sides: if both the diagonal and the sides are composed of a certain number of indivisible parts, then one of these indivisible parts would constitute a common measure of the two line segments. Examples of the second type of proof involve constructions that aim to establish infinite divisibility from infinite extendibility, for example, one is asked to imagine a ship that is sailing off in an infinitely extended flat sea. As the ship sails away, its apparent size becomes infinitely small. A combination of these two types, where a construction is used as a basis for a *reductio* is also prevalent in these texts, for example, two concentric circles are used to demonstrate how from the assumption of finite divisibility one derives that the number of points on the circumference of each circle is the same, from which the absurdity that their circumferences are equal in size follows.

We shall not pause to discuss these proofs here, as they were conceived before the development of the modern, set-theoretic, notion of infinity.[3] But in the nineteenth century, with the works of Cauchy, Dedekind, Weierstrass, and Cantor, the notions of limit and convergence became rigorously formalized, and infinities were re-conceived within the axioms of set theory. In this context the line segment is

---

[2] Most of these demonstrations are found in *The Port-Royal Logic* [Bay51], a logic textbook first published anonymously in 1662, and in Isaac Barrow's *Lectiones Mathematicae*, first published in Latin in 1683.

[3] The above arguments were considered suspect by philosophers such as Hume, Reid, and, more famously, Berkeley, who had already criticized infinitesimals and the shaky foundations of the seventeenth century calculus in *The Analyst* [Fog88, Cum90].

postulated *ab initio* to be a continuum, regardless of any actual infinite process by which it is assumed to be constructed. The first horn of the dilemma in Zeno's paradox of extension – about the convergence of unextended parts – could now be analyzed with greater care, and modern mathematics could once again demonstrate its consistency.

2.2.3 A dissolution

This task, carried out by Adolf Grünbaum in his doctoral dissertation [Grü52], consists of several steps. First, to evaluate Zeno's assertion that a finitely extended line cannot be regarded as an aggregate of points, no matter what cardinality one postulates for the aggregate, we need to clarify the logical relationship between the modern concepts of cardinality, dimension, metric, length, and measure. In particular we need to disambiguate the notion of "extension," which intuitively has been interpreted interchangeably as "dimension" or "length."

First let us interpret "extension" topologically as "dimension." In this case, the appropriate context for evaluating Zeno's dilemma is dimension theory. In this theory, the existence of sum theorems shows that the line, defined as one-dimensional, can be decomposed to aggregates of points (rational and irrational), all of which have dimension 0, and so the sum of zero-dimensional sets need not be zero-dimensional ([HW41], p. 18).

On the other hand, "extension" may be interpreted metrically as "length." In this case the arena for evaluating Zeno's assertion is measure theory and the linear Cantorian continuum, where the following definitions hold.

(1) A finite interval on a straight line (a one-dimensional Euclidean space) is the (ordered) set of all real points between (and sometimes including one or both of) two fixed points called "end-points" of the interval.
(2) The length of the points-set constituting the interval (a, b) is defined as the Euclidean distance between the interval's fixed end-points $b - a$.[4]
(3) A *degenerate* interval (whose end-points coincide, i.e., $a = b$) is a point (or more precisely, a singleton whose member is a point), and is of length 0.

Next, length, and its generalization to points-sets other than an interval, namely, measure, are *countably additive*. To calculate the length of an interval composed of subintervals we use the standard definition of an arithmetic sum: if a non-degenerate

---

[4] In general, the Euclidean distance function between the points (x_i, y_i) in Euclidean space of dimension n is given by

$$d(x, y) = \left(\sum_1^n (x_i - y_i)^2 \right)^{1/2}. \tag{2.1}$$

interval is composed of subintervals of finite cardinality, no two of which have a common point, then its length is equal to the arithmetic sum of the individual lengths of the subintervals. Generalizing the definition of arithmetic sum to the countably infinite domain, we can define it as the limit of the sequence of partial arithmetic sums of members of the sequence. This allows us to equate the length of an interval composed of subintervals whose cardinality is *countably infinite* with the arithmetic sum of the lengths of these subintervals.

But here's the rub: if an interval is composed of subintervals of *uncountable* cardinality, these subintervals *must* be degenerate, and the notion of an arithmetic sum ceases to hold. In other words, the assignment of length in this case does *not* depend on the arithmetic sum (or any extension thereof)! In addition, length and cardinality, while both being properties of sets, are nevertheless logically independent: the cardinality of two points-sets may be the same, but their length may differ.[5]

This independence of cardinality and length, and the fact that the real line is uncountably infinite, together block Zeno's argument from premises (2) and (3) above to the paradoxical conclusion that the length of a positive, non-degenerate interval is 0. Consequently, measure theory and Cantor's analysis of the uncountable infinite allow us to affirm consistently and simultaneously the following four propositions: (a) the finite interval (a, b) is the union of a continuum of degenerate (sub)intervals; (b) the length of each degenerate (sub)interval is 0; (c) the length of the interval (a, b) is given by the number $b - a$; (d) the length of the interval is not a function of its cardinality.

We see that in both the *topological* and the *metrical* sense, modern mathematics survives Zeno's paradox of extension. But apart from restoring consistency to the notion of the continuum, the above analysis leaves us with an important lesson. The refutation of Zeno's paradox depends crucially *both* on the notion of countable additivity *and* on the structure of the line segment. Since, according to Grünbaum, there are good independent reasons to maintain the former (as it is instrumental in the foundations of analysis, geometry, and statistics), in order to avoid contradictions, the latter must be of cardinality of the continuum, i.e., $\aleph_1$. Otherwise, since any denumerable set (e.g., the rationals) is of measure 0, paradox will ensue.

Grünbaum concludes that given the wide applicability of the notion of countable additivity in applied mathematics, any consistent structure of the line segment must be either discrete (i.e., finitely divisible) or infinitely divisible but with cardinality $\aleph_1$. What the modern discussion on Zeno's paradox of extension has taught us, says Grünbaum, is that a geometry which is based on a countably infinite line segment

---

[5] Thus the largest three-dimensional region contains the same number of points as the smallest line segment, and even for a non-measurable set of points in, say, a sphere, there will be a corresponding part of the sphere which exactly fills that set of points. The latter result is known as the Banach–Tarski paradox (see, e.g., [Wag93]).

(with cardinality $\aleph_0$), such as the cardinality of the rationals or the integers, should be rejected on logical grounds alone.

Grünbaum uses this result to criticize Russell, who famously supported a geometry based on a line segment with cardinality $\aleph_0$,[6] as well as other attempts, for example, Whitehead's, to construct a geometry based on finite yet unbounded, discrete operations.

A more appropriate lesson, to my mind, is that the modern discussion has exposed the price of a geometry that purports to describe physical space and which relies on a line segment with cardinality of $\aleph_0$ (see below, Section 2.4): a geometry of this sort inevitably leads to a mathematical system in which countable additivity ceases to hold at the most fundamental level.

2.3 Topology and the argument against collision

2.3.1 Background

Debates on actual infinite divisibility appear also in discussions on the nature of extended substances and their boundaries. Historically, questions regarding the boundaries of three-dimensional objects (e.g., "are three-dimensional solids bounded by two-dimensional surfaces, surfaces by one-dimensional edges, and edges by dimensionless points?," or "where does a sphere touch a flat object if it has no unextended parts?"), together with Poincaré's purely topological account of the notion of dimension, served as basis for the development of topology at the end of the nineteenth century. The debate that fueled this development is worth mentioning, as it harbors some of the best attempts to argue for and against actual infinite divisibility on *a priori* grounds.

There are, broadly speaking, three doctrines about physical boundaries that can be found in the medieval and modern debates (see [Zim96]): indivisiblism, moderate indivisiblism and anti-indivisiblism. The first, in its extreme, amounts to recognizing only indivisible physical substances. The last amounts to denying that extended objects possess any less-than-three-dimensional parts. The moderate view admits the existence of indivisibles (point, edge, surface), but also accepts Aristotle's idea that extended objects cannot be composed of indivisibles alone. It consequently recognizes two kinds of parts in every extended body: infinitely divisible three-dimensional stuff, and physical indivisibles that terminate and connect the object's three-dimensional parts.

---

[6] Russell saw such a space as indistinguishable from the continuum:

> A space … in which the points of a line form a series ordinally similar to the rationals, will, with suitable axioms, be empirically indistinguishable from a continuous space, and may be actual.
>
> *([Rus38], p. 444)*

These three doctrines survived throughout the seventeenth, eighteenth, and nineteenth centuries, with notable advocates: Berkeley and Hume built extended entities out of a finite set of unextended elements ("minima sensibilia"), promoting indivisiblism. Bolzano followed them, yet for him the above set of unextended elements was infinite. Descartes, Malbranche, and Euler, on the other hand, defended anti-indivisiblism, and Brentano supported moderate-indivisiblism. The details of the reasons given in each case are unimportant to our discussion, but the context is. For the attempts to discern the structure of material bodies (and, indirectly, the structure of space) using *a priori* reasoning have resulted in an interesting argument [KM87] that, as in the case of Zeno's paradox of extension, can be used to challenge the coherence of the notion of the continuum as an aggregate of dimensionless points.

2.3.2 *The argument against collision*

Material bodies are extended, i.e., they occupy regions of space and, presumably, are able to move and to collide with each other.[7] Without loss of generality, we can limit our discussion here to motion in one dimension. Suppose the spatial region, occupied by a material body, is topologically closed; each body thus occupies the boundary points of its volume. In order for two bodies to collide, or to be in contact without overlapping, their end-points need to be adjacent. But if space is a continuum, no two points are ever adjacent to each other. Suppose now that the spatial region is topologically open, i.e., material bodies occupy open intervals, and no body occupies the boundary points of its volume. But now, no matter how close we push two bodies together without their being overlapping, the completeness of the linear continuum entails that there will always be a residual, unoccupied, point between them, so contact is again impossible.

Assuming that the argument is valid, one can resist its conclusion by denying one or more of its premises. Denying that space is finitely divisible and discrete is but one of the possible ways of doing so. One can also deny that material bodies do not overlap, or claim that matter is composed of indivisible topologically *open* atoms (since a finite collection of topologically open objects is itself topologically open). One may even hold a view such as Bolzano's (e.g., [Bol50], pp. 167–168) in which bodies are only topologically half-open, occupying, say, the right but not left end-points of any spatial interval. On this view collision would be possible if closed-sides of bodies only encountered open-sides of bodies. While definitely consistent, this view is somewhat contrived, especially when we generalize our one-dimensional case to collisions in three dimensions.

---

[7] While this description is admittedly substantival, what follows does not hinge on spacetime substantivalism. The same argument can be made from a spacetime relationalist perspective.

An altogether different response, assuming, again, that the argument is valid, might be to "bite-the-bullet" and to agree that no contact or collision between material bodies is possible, and, assuming no action-at-a-distance, to declare that, independently of any *empirical* evidence, field theory is true as it provides a collision-friendly alternative to atomistic mechanics.

In field theory the naive picture of two billiard balls colliding can be described in such a way that eschews the problems generated by the *a priori* claims about what contact requires:

> After we have introduced fields of force, we may be able to picture the 'collision' between two balls as involving repulsive fields driving the balls apart. Each ball interacts only with the field 'surrounding' the other ball. A 'collision', then, is just a close approach between two balls – close enough that each ball penetrates deeply into the other's field, to where that field becomes very strong. Spatiotemporal locality is guaranteed by each ball's interacting only with the field at its location.
>
> *([Lan02], pp. 12–13)*

Here the balls do not interact directly with each other. Rather, they interact with the field produced by the other, while each is located *in* the other's field. The contact here is thus between the balls and what produces the force on them, namely, the field. The latter is allowed to penetrate into them, and so in this way we have contact and we need not worry about penetration.

But is the argument against collision valid? One attempt to deny its validity is to note [Haz90], that the distinction it rests on, between closed and open intervals, is *in principle* unascertainable with any physical measurement process, i.e., in order to make this distinction one would be required to attain actual infinite precision (and not just finite but unbounded precision) in a measurement process. But a dynamical process that requires actual infinite precision is also indeterministic (see Section 2.6), and so the price for the argument's validity would be denying our physical theories one of their essential features. A counter-attempt to forfeit this price, by saying that the argument need not depend on our ability to discern the above distinction between closed and open intervals, is self-refuting: after all, the argument's conclusion is some feature of the physical world, the only access to which is given by our scientific theories, and the verification and confirmation of these do depend on our measurement capacities.

Another, related, attempt to deny the argument's validity is to deny that the (perfectly consistent) topological analysis it employs has any bearings on the actual structure of physical space. The idea here is that while the set of real numbers represents the structure of physical space, it does so only abstractly: each real number represents, not a concrete spatial entity, but a series of outcomes of a possible (infinite) measurement process. This mode of thought goes back to

Whitehead's "method of extensive abstraction," which seeks to disregard points as geometrical primitives, and to describe them instead by showing how the geometric relations between points emerge from the ultimate relations between the ultimate things which are the immediate objects of perception (see, e.g., [Whi19], p. 5).

On Whitehead's view, actual infinity can never be perceived, hence points are not genuine natural entities. Nevertheless, they can be reconstructed from sense perception via an infinite sequence of nested spatial intervals with – for any given length – all but finitely many intervals of the sequence being less than that length. Any such sequence *covers* another if every interval of the first contains intervals of the second. The covering relation is reflexive and transitive, and two sequences are thus equivalent if they bear to each other the symmetric and transitive closure of the covering relation. Whitehead's points are thus equivalence classes of these sequences under the above equivalence relation.[8] On this view, the end-point of any interval is an abstract entity and, as such, should not be construed as a concrete part of the extended body. Note that this view is empirically indistinguishable from the view in which the end-points are part of the extended body, and so the distinction between open and closed objects upon which the argument against collision rests turns out to be not a physically meaningful one. Whether or not bodies occupy their end-points becomes a matter of convention, and unless points are (mis)construed as physical parts of space, neither consequence is problematic.

Adolf Grünbaum criticized Whitehead's method [Grü53] by pointing out that it fails to satisfy the constraint imposed by Zeno's paradox of extension. The consequence of Grünbaum's analysis of this paradox, recall, is that if infinite divisibility is allowed, then a consistent structure of the line interval must have cardinality $\aleph_1$. Since sense awareness cannot suggest an interval structure whose cardinality exceeds $\aleph_0$, it follows that if the actual structure of the interval is $\aleph_1$, then sense awareness cannot perceive it. And since one of the necessary conditions of Whitehead's method is that points must have to each other the kind of relations which geometry demands, then if space does form an actual continuum of cardinality $\aleph_1$, Whitehead's method fails.

The failure of Whitehead's method is also evident, says Grünbaum ([Grü53], pp. 219–220), if one considers that it cannot disambiguate two irrationals, as required by Dedekind's definition of the term. For if sense perception is the sole generator of points, then there is no way to distinguish between, say, $x = 0$ and $x = 10^{-1000}$. This difficulty is intrinsic to the positivistic assumptions on which Whitehead's method rests, and issues in Whitehead's failure to provide for distinguishing a given point from a continuum of others.

---

[8] This definition, given in [Haz90], p. 209, is slightly different than the one originally given by Whitehead, but is sufficient for our discussion. For an application of this method in a construction of metric spaces based on primitives such as three-dimensional solids see, e.g., [GV85].

Upon reading Whitehead more carefully, it appears that Grünbaum may have been barking at the wrong tree. As one recalls, the lesson from Zeno's paradox is not that the only consistent structure of the real line is of cardinality $\aleph_1$. Another option, namely, finite divisibility, or discreteness, is as consistent. Whitehead could have been making exactly this choice. On many occasions he stresses ([Fit79], p. 136) that while continuity is potential, actuality is only discrete:

It cannot be clearly understood that some chief notions of European thought were framed under the influence of a misapprehension, only partially corrected by the scientific progress of last century. This mistake consists in the confusion of mere potentiality with actuality. Continuity concerns what is potential, whereas actuality is incurably atomic.

([Whi29], p. 61)

This view is also expressed succinctly in Hilbert's "On the Infinite":

A homogeneous continuum which admits of the sort of divisibility needed to realize the infinitely small is nowhere to be found in reality. The infinite divisibility of a continuum is an operation which exists only in thought. It is merely an idea which is in fact impugned by the results of our observations of nature and of our physical and chemical experiments.

([Hil25], p. 186)

So the real dispute between Grünbaum and Whitehead should be construed not as a dispute about logic but as a dispute about reference. The issue at stake should not be the consistency of the two alternatives to the line segment, the continuum and the discrete, but rather their applicability to physical reality. Grünbaum's objection amounts to saying that these two consistent alternatives are mutually exclusive in their applicability to physical reality, so that there is no way even to approximate the continuum with the discrete. According to Grünbaum, if physical space forms a continuum of uncountably infinite dimensionless points, then no discrete geometry can describe it correctly, and vice versa. We shall return to this thesis in Section 2.4.

2.3.3 Lessons from continuum mechanics

In our analysis of the argument against collision we have so far relied on the pre-theoretic and *a priori* conceptions it employs of "contact" or "collision." Since the argument purports to derive contingent conclusions about the nature of the physical world from these conceptions, it is quite rational to raise doubt with respect to such pre-theoretic notions. A short detour to physics may thus prove instructive.

We have already seen that field theory can describe collision without actual contact. But can we consistently describe collision *with* contact in a continuous background? It turns out that we can. One such option, described thoroughly in [Smi07], is continuum mechanics. This sophisticated branch of mathematical

physics was initiated primarily by Euler and Cauchy as a generalization of Newtonian mechanics, flourished in the late nineteenth century, and was pursued by a
great number of the leading scientists of the time including Lord Kelvin, Stokes,
Maxwell, and Hadamard.

Admittedly, the microstructure that real-world bodies have is quantum mechanical, and not the structure attributed by continuum mechanics; bodies are not continuous, and their interaction is not described with classical trajectories. Nonetheless,
continuum mechanics is still an active area of research with many applications in
applied math and engineering, for example, fluid flow, gas dynamics, and deformation of elastic and plastic bodies. But as in the debate between Grünbaum and
Whitehead, one should distinguish between problems of consistency (logic) and
problems of truth (reference). Denying the validity of the argument against collision belongs to the former domain, and so a consistent physical theory of the
continuum in which collision *is* possible is sufficient for this purpose; it matters
little whether this theory is actually *true*, i.e., has any reference, or truth makers, in
the real world.

Within continuum mechanics the notion of impenetrability (crucial for the above
argument against collision) is kinematically defined as a uniqueness constraint on
the mapping between the space of configuration of material bodies and physical
space: no material point can occupy two spatial points, and no spatial point can be
occupied by two material points ([LS74], p. 419). But kinematical constraints are
often relaxed by dynamical considerations in applied mathematics,[9] and there is
no way to know in advance when such relaxation is called for, especially when one
models dynamics in a continuous background. When two material bodies come
into contact in continuum mechanics, a violation of the above constraint is required
at exactly the boundary of the material body, i.e., on a set of Lebesgue measure
zero ([Smi07], pp. 511–516).

Of course one could argue that any violation of the kinematical constraint above
is a violation of impenetrability, but it is not clear on what grounds this claim could
be defended. After all, applicability up to a set of Lebesgue measure zero (i.e.,
in *almost all* cases) is not rare in mathematical physics, for example, in statistical
mechanics or in real analysis, and it is often the case that a behavior on that set is
quite different from the behavior on the set of positive measure. As in many other
cases, our pre-theoretic notion of impenetrability may as well require revisions in
light of physics, and not vice versa.

Another pre-theoretic notion that may require revision is the notion of "contact"
([Smi07], pp. 520–524). Here also, the idea is that the appropriate definition of
contact between continuous bodies should be driven primarily by physics and the

---

[9] For example, the kinematical requirement for twice-differentiability may be relaxed in certain cases of non-linear
   partial differential equations with singularities.

applied mathematics that govern the dynamics, and could not even be guessed *a priori*.

The basic building block behind the definition of contact in continuum mechanics is the notion of "contact force," which is quite different than the notion of a "body force": the contact force is a surface integral acting only on the boundary of the body, whereas the body force is a volume integral acting throughout the interior. Also, if we were to increase the mass of a finite portion of the interior of the body, the total body force would change, because it is sensitive to that alteration of the interior, but the total contact force would not, as long as the boundary of the body remained the same. Moreover, contact forces are constrained dynamically in such a way that forbids action-at-a-distance and restricts the density of the contact force on the boundary to what is happening in an arbitrarily small neighborhood of the boundary, so that neither the body forces nor the far exterior of the body can influence them.

One consequence of continuum mechanics and the forces it defines is that a point cannot be regarded as a material body since a point has no boundary, and so the notion of a contact force does not make sense in this case. In continuum mechanics one starts with the regularly open sets as the class of bodies, in part so as to exclude the individual points as bodies! Note that this result follows from physical considerations, and not from positivistic ones such as those promoted by Whitehead.[10]

What follows from the definition of contact force and additional mathematical considerations (see [Smi07], pp. 525–527) is a physical definition of "contact" that will allow us to make precise our pre-theoretic notion: two bodies are in contact if they have a non-empty contact set, where the contact set is the set of points along which a contact force between two bodies can be present, and is given by the intersection of the reduced boundaries of the two bodies.[11]

Clearly, and contrary to the assumption of the argument against collision, this definition allows contact between open bodies (bodies which do not contain their boundaries) since their reduced boundaries may still intersect.

Equipped with a physical definition of "contact," we can now compare it to possible pre-theoretic definitions (see [Smi07], pp. 528–531). One such possible pre-theoretic definition is that two bodies are in physical contact if and only if there is zero distance between them. Here, as in the case of Zeno's paradox of extension, bodies are conceived as sets of points, and so the notion of distance needs to be disambiguated.

---

[10] The latter may be "naturalized" by using the former: measurement requires interaction, and if a point has no boundary, then no force can be defined to resolve it in a measurement.

[11] The reduced boundary is a finite area measure subset of the topological boundary where an outer norm for the contact force can be defined.

If by "distance" one means *metrical* distance, where the distance between a point set X and a point set Y in a metric space is (by definition) the greatest lower bound of the set $\{d(x, y) : x \in X, y \in Y\}$ where $d(x, y)$ is the distance between x and y, then, at least according to continuum mechanics, two open sets (i.e., material bodies) could be in contact without overlapping.

If by "distance" one means *topological* distance, where contact is defined as topological *connectedness*, then continuum mechanics is of no avail, but then, as in the case of "impenetrability," what independent grounds does one have to insist on defining "contact" this way? Surely this is not the most intuitive definition, and even if it were, it is not clear why considerations from physics – such as the one brought here – cannot reshape it.

The upshot is that continuum mechanics defines the basic notions of "contact" and "impenetrability" consistently within a continuous background,[12] and in so doing supplies us with counterexamples to the argument against collision, militating against its validity. To repeat, it may well be that, in the ultimate theory, matter is not continuous but, as argued above, this fact (at least at this stage of our discussion) is completely besides the point.

What we have learned here, again, is that *a priori* arguments for (or against) the continuum as a consistent description of the line segment are suspect. The real issue at stake is not the formal consistency of these descriptions, but their applicability as a description of physical space.

2.4 Geometry and the tile argument

2.4.1 Discrete geometry

The history of mathematics has seen several attempts to construct a geometry based on a discrete line segment. The motivations for such constructions are diverse, and span perspectives such as intuitionism, finitism, and mathematical empiricism, and we shall say more on these in Chapter 3. The basic idea is to regard geometry as a formal system, generated by finite means by a geometer, whose capacities are bounded and finite. On this view, geometry is more an experimental science than a branch of pure mathematics.

Prima facie discrete geometry may appear less mathematically tractable than its continuous counterpart, but it actually turns out to be more simple: one can, for example, characterize the whole of the discrete geometry in terms of a single symmetric and irreflexive dyadic relation of *adjacency* (having no points in between) – where "distance" is defined as the number of adjacent points – which

---

[12] Note that in field theory the notion of contact is also defined with forces, yet these forces are body forces, while in continuum mechanics they are contact forces (see [Smi07], pp. 533–535).

seems quite elegant.[13] In contrast, the continuous structure may require a distance function which is at best a five-place relation, in which four of the relata are points and one is a number. To express this number in terms of points we can use Hilbert's axioms for Euclidean geometry, but these require two primitive relations – a triadic *betweenness* and a tetradic *congruence*. Luckily for the supporters of the continuum, however, elegance and convenience need not be taken seriously as criteria for deciding the contingent character of physical space.

In constructing a discrete geometry one can either (1) make changes to the axiomatization of the Euclidean plane (as laid out by Hilbert in 1899), so that finite models can satisfy the modified axioms, or (2) keep the axioms intact, but construct finite models that approximate the infinite ones as closely as possible. Attempts to construct finite models that approximate the standard continuum geometry were first made in the 1940s and 1950s by a group of mathematicians based in Helsinki. Their work, inspired by an earlier treatment by J. T. Hjelmslev from 1923, sought to replace the mathematical representation of the infinite plane as a mapping to the Cartesian product of the real numbers, with a mapping from the plane to a finite number field (see [VB10] for more details). Early on it became clear that this replacement was not problem-free: for example, not all finite fields will do; the admissibility depends on the *size* of the field.

The basic intuition of the Helsinki group was to focus on finite fields, also called Galois fields, which are all classified and are simply of the type $GF(p^n)$ (where $GF(p^n)$ is the field of integers $\mathbb{Z} \bmod(p^n)$, p is a prime, and $n \in \mathbb{Z}$), and to find such a finite field, a sufficiently large portion of which is "like the real number system" with which one could describe the observable universe, i.e., a range between 10^{-13} and 10^{27} (the range between one Fermi and the distance to the farthest known object in the universe). Clearly there is no difficulty in finding enough points from a field $GF(p)$ provided p is large enough; it has been shown that to fill this range a prime $p \sim 10^{10^{81}}$ is sufficient, see, for example, [Coi59].

But to approximate the real number system, such a subset of the finite field, no matter how huge, must also be transitively ordered, a very non-trivial constraint given the periodicity of the finite field. And yet, what the studies from the 1950s showed was that if the prime is chosen to have the form[14]

$$p = \left(8x \prod_{i=1}^{k} q_i - 1 \right), \qquad (2.2)$$

---

[13] Note also that this primitive notion also *explains* what in the continuum case is taken to be an *unexplained* axiom, namely, the triangle inequality (that the distance from P_1 to P_3 cannot exceed the sum of the distances from P_1 to P_2 and P_2 to P_3).

[14] The existence of a prime of this form is guaranteed by Dirichlet's theorem, that states that for any two positive co-prime integers a and d, there are infinitely many primes of the form $a + nd$, where n is a non-negative integer. In other words, there are infinitely many primes which are congruent to a mod d.

where x is an odd integer and $\prod_i q_i$ is the product of the first k-odd primes, then -1_q is "negative" and 2 and the first k-odd primes are "positive." For such a prime the first N integers for large N can be (locally) transitively ordered and consequently the geometry in that neighborhood would appear to be like the ordinary Euclidean plane, up to very large (and down to very small) distances [Mor74].

Later work [RS69] incorporated these abstract considerations from number theory into physics by developing concepts such as order, norm, metric, and inner product over the above subset of the total finite field in which transitive order could be defined. With these "extensions" it became clear that a finite discrete space behaves locally (albeit not globally) like the standard conventional continuum. This insight, namely that a discrete description of physical phenomena in the neighborhood of the "ordered" subset of the total field is locally indistinguishable from the standard continuum description, was also repeated by Schwinger ([Sch01], p. 84), and has recently reappeared in attempts to approximate the continuum of the Hilbert space (as a vector space over the complex numbers) with a vector space constructed over a specific Galois field $GF(p^2)$ of the sort described above [HOST13b].[15]

A stab at discrete geometry of the second kind was taken by the philosopher Patrick Suppes [Sup01], who urges us to view geometry as a practical science which uses only finite notions and definitions. A formulation of this sort is completely consistent with Greek geometry. The shift to non-finitistic methods, says Suppes, entered classical analysis and was extended, in some sense, to geometry, at a certain point in the nineteenth century. The key motivation for this shift was that representations of the actual infinite were required for a structure-preserving isomorphism between the geometrical plane and the Cartesian product. The issue is but a part of the more general question of the use of actual infinity in the sciences.[16] Suppes' own opinion is succinctly put:

It is a separate question to what extent the parts of classical analysis actually used in applications of mathematics in physics and elsewhere can avoid commitments not only to any results using the axiom of choice, but, also, any sets of an actually infinite character. My own conviction is that one can go the entire distance, or certainly almost the entire distance, in a purely finitistic way . . .

([Sup01], p. 136)

Faithful to this conviction, Suppes introduces two basic operations, namely, bisecting a line and doubling a line. The basic building block is then defined as a step in the construction which consists of three elements: the (new) point to be constructed,

---

[15] Insofar as this vector space can approximate (locally) the notion of an inner product, and can support showcase quantum algorithms [HOST13a], these attempts have also succeeded in reproducing the empirical content of non-relativistic quantum mechanics from an underlying finite and discrete structure.

[16] See [Arn12] for a general discussion on this question.

a pair of points already present, and one of the two basic operations (bisecting or doubling). Using such basic steps, one can construct, say, a parallelogram, which then allows one to initiate a formal axiomatic treatment (e.g., listing the necessary axioms that the basic operations must satisfy for the construction to be counted as a parallelogram, and proving a representation theorem that maps the constructed points to rational coordinates).

An even more ambitious attempt that maps geometrical points to integers was presented in the work of the Polish-American physicist Ludwig Silberstein [Sil36]. Admittedly, Silberstein discusses only one spatial dimension, and so in this particular case the Euclidean and the discrete distance functions coincide. Differential equations are replaced with difference equations, and finite differences are derived with Taylor series expansions. The construction is rich enough to derive most of classical physics. Silberstein goes further to apply it to the special theory of relativity, suggesting a discrete spacetime analog, which, while unknown to most working physicists in Europe and the USA at that time, had a certain influence on the attempts of the late 1940s to establish a quantum mechanics based on discrete spacetime (see Chapter 5). In this context he remarks:

When the dimensions are atomic or subatomic and the time intervals implied of the order of $10^{-18}s$, the Principle [of special relativity] may, according to our theory, break down. But can we claim to know that it does hold in the atomic or subatomic world? Of course not. All modern physicists are inclined to believe that our usual, molar physics, including our space and time concepts, are inapplicable in such circumstances.

([Sil36], p. 41)

2.4.2 Weyl's tiles

The above attempts are, of course, the exception and not the rule. Geometry applied in the physical sciences is customarily continuous; the use of the real number system to represent spatial (and temporal) physical magnitudes is characteristic to modern as well as classical physics. To be regarded as a serious alternative, a discrete geometry – assuming it is consistent – must at least reproduce the success of this characteristic applicability. In other words, any consistent geometry based on spatial discreteness that purports to describe physical space at the small scale (where "small" should be rigorously defined via some physical ratio or magnitude), must also reduce to, or at least approximate, a continuous geometry at the large scale. The required relation between the two geometries here is similar to the relation between Euclidean and non-Euclidean geometries: while Einstein's general theory of relativity shows that the latter is the correct description of the properties of physical space, it also shows that Euclidean geometry is a good approximation to these properties where the gravitational field is not too strong.

Modern physics in its current state is unable to decide this issue *in practice*. So far there is no agreed upon theory of space at the small scale, and even if there were such a theory, it is not clear what phenomenological evidence could indicate its applicability (more on that in Chapters 7 and 8), but throughout the history of math, several arguments have appeared whose aim was to demonstrate that the required approximation cannot be achieved even *in principle*.

Among these arguments, the most famous is Herman Weyl's tile argument:

So far, the atomistic theory of space has always remained mere speculation and has never achieved sufficient contact with reality. How should one understand the metric relations of space on the basis of this idea? If a square is built up of miniature tiles, then there are as many tiles along the diagonal as there are along the sides; thus the diagonal should be equal in length to the side.

([Wey49], p. 43)

Many (e.g., Salmon [Sal80], pp. 65–66) see Weyl's argument as decisive against the ability even to approximate the obvious features of macroscopic space with a discrete alternative. For no matter how small Weyl's tiles may be made, Pythagoras' theorem (which is at least approximately true in physical space, as we have found by much experience) is always violated – the proportionate error, namely $\sqrt{2}$, does not decrease as the number of tiles increases, and so the Euclidean distance function (equation (2.1)) cannot even be defined.

Can the tile argument be resisted? First, one may note that in a discrete space, "length," or "distance," is given by the number of contiguous building blocks between any two end-points:

To construct any such concept of distance we must first tie up in some way or other the elements of different strings of elements, i.e., one-dimensional paths, by introducing an enriched concept of neighbor which connect any two points in the discrete space (logically, as an undefined term), by giving each element a certain, say fixed number m of neighbors, and thus providing for a possible passage from every one to every other element of the manifold.

([Sil24], p. 363)

The tile argument targets discrete geometries whose distance function is based on a structure of eight elements contiguous to each element. One can show (e.g., [Rog68], p. 122) that when the distance function is constructed on the basis of a different discrete structure, say a structure of six elements contiguous to an element, or hexagons, it can approximate the macroscopic Euclidean distance up to the nearest integer at least sometimes, but will fail to hold for certain privileged directions relative to the grid. This may indicate that the tile argument is dependent on a particular structure of the discrete geometry, and that a further study of different abstract discrete structures may lead to one that can be given a physical interpretation free of the problem pointed out by Weyl.

Another attempt to blunt the tile argument introduces constant width to the line segment in the discrete structure [VB87] so that the length of the line segment is given by the sum of the tiles it is made of *modulo* its width. The tile argument targets the limiting case in which this width is always 1, but once a non-trivial width is admitted, Pythagoras' theorem can be again approximated to any degree of accuracy as the tile structure becomes smaller and smaller.

The problem with this solution is that the lengths of the sides of a triangle as well as the length of the hypotenuse are determined with considerations from Euclidean geometry, and so one cannot say here that the latter is approximated by a structure more fundamental. Consequently, while this solution shows that the metrical relationships of classical Euclidean geometry *can* be related to discrete relationships in the tile-structure (which is precisely what Weyl claims to be impossible), it still falls short of fulfilling the more ambitious aim of constructing the former out of the latter without any reference to it.

A third attempt to resist the tile argument is to note that it saddles the candidate discrete structure with unnecessary constraints which can be relaxed without compromising the admissibility of the said structure [For95]. If the tile argument is taken literally, it requires points to have extension as they are identified with tiles. This, in turn, seems to imply that part of a point is some distance from some other part of the point, which contradicts the premise that there is no distance smaller than the distance between two adjacent points. Since parts of a partless square are as nonsensical as a squared circle, instead of taking the tile argument literally, we should see it as *one* way of representing a possible discrete geometry using Euclidean space. What the argument then shows is that adjacent points of an admissible discrete geometry cannot be represented in Euclidean space as extended regions, but why should an admissible discrete geometry be represented in this way to begin with? Agreed, such a representation is easy, but we have already seen that ease of representation is not a serious requirement.[17]

I therefore conclude that, contrary to the received wisdom, Weyl's tile argument need not forestall us from considering discrete geometries and their putative ability to approximate their continuous counterparts in the macroscopic scale.

2.4.3 Zeno's stadium

Another famous argument against the *in principle* ability to approximate continuous geometry with a discrete one is Zeno's stadium:[18] consider three rows of objects, *A*, *B*, and *C*, arranged as indicated in Figure 2.1 on the left. While row *A* remains

---

[17] Forrest ([For95], pp. 331–333, 341–352) describes a possible representation of discrete geometry based on the notion of adjacency which does not fall prey to the tile argument, and is also independent of features of the continuous counterpart – in its symmetries and in its notion of dimension – hence is not parasitic on the latter.

[18] This reading of the paradox is due to Russell, see [Rus29], pp. 182–198.

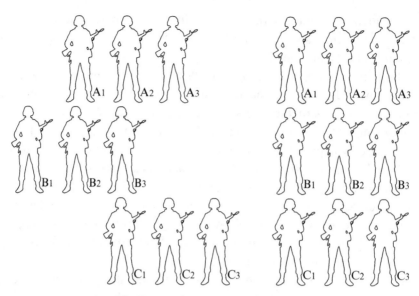

Figure 2.1 Zeno's stadium paradox.

at rest, rows B and C move in opposite directions as to attain the arrangement on the right.

In this process, C_1 passes twice as many Bs as As. A contradiction then follows if one assumes that a finite line interval contains a *finite* number of points, and is dissolved if one assumes that between any two points there is another point. But, as has been pointed out ([Rog68], pp. 119–121), what one requires from a certain admissible discrete microstructure in this context is not that its properties should mimic exactly the properties of the macrostructure; rather, the only requirement is that the microstructure account for the macrophenomena. And so, while the motion of macro-objects (Russell's example is a drill sergeant and his troops) seems to obey the premise that between any point there is another (and so if an object is approaching another at one moment, and is beyond the other object at another moment, then there must be a moment at which the two objects are facing each other), the motion of the microconstituents need not obey the same premise.

2.4.4 Upshot

Other problems with discrete geometry have to do with its applicability *in practice* to modern physics. For example, in a discrete geometry of spacetime, continuous motion is hard to define; hard, but not impossible. While an upper limit to velocity is a natural consequence of this structure, velocities below the upper limit can be accommodated with a "jerky" motion, where the particle remains still for a time at a certain place and then moves.[19] Also, a discrete spacetime must admit privileged directions, depending on its structure, but this does not mean that at the macroscopic scale the anisotropy will vanish.[20] Finally, the possibility of a relativistic discrete spacetime is also a matter of dispute. We shall pick up this issue in Chapters 5–8.

As in the case of topology, the upshot is that so far there exists no "no-go" argument – based solely on logic – that can establish the impossibility of approximating continuous geometrical structures with discrete ones. We are thus left with two options regarding the question which structure is more fundamental, i.e., which structure actually applies to the physical world: either it is an *empirical* question, or it is a matter of convention. What is clear is that one cannot settle the issue with considerations from logic alone.

2.5 Richer alternatives to the real line model

2.5.1 Background

Up until now we have been discussing arguments for and against attempts to construct discrete geometries whose coordinate systems can be mapped to either countably finite or strictly finite substructures of the real line segment. In the 1960s, however, there appeared richer alternatives to the real line segment, which opened up the question whether these alternatives could serve as a basis for different geometries.

From a strictly formal perspective, the properties of the real line model satisfy the 11 axioms of a complete ordered field.[21] Among the consequences of model theory and Gödel's incompleteness theorems is the one that says that this satisfaction is non-unique, i.e., there exist many – in fact, infinitely many – models that satisfy these axioms. One such model, constructed in 1963 by Paul Cohen, can also be used to represent discrete spatial geometry.

The model results from Cohen's celebrated proof of the logical independence of Cantor's continuum hypothesis (CH henceforth) of the ordinary (Zermelo–Frenkel) axioms of set theory. The CH amounts to the claim that every infinite set of real numbers is either countable (with cardinality $\aleph_0$) or uncountable (with cardinality $\aleph_1$, or c), so that there are no sets of intermediate size.[22]

---

[19] Russell was aware of this problem ([Rus27], p. 375). For a possible solution see [Sil36] and [For95], pp. 337–338.

[20] See [For95], pp. 338–339, and [VB10].

[21] These axioms are *Addition and Multiplication, Associativity, Commutativity, Distributivity, Identity, Additive Inverse, Multiplicative Inverse, Trichotomy, Transitivity, Isotony,* and *Completion*. The first 7 axioms define a field; the first 10 axioms define an ordered field, see, e.g., [Lyn71].

[22] Cantor formulated the CH without proving it in the nineteenth century, and it was listed as one of Hilbert's celebrated 23 problems for mathematics at the dawn of the twentieth century.

The objects in the new model are real-valued functions defined on some set S, i.e., $f : S \rightarrow \mathbb{R}$,[23] where R^S is the set of all functions f from S to $\mathbb{R}$. A quick look shows that R^S is not even a field, as it fails to satisfy *Multiplicative Inverse*.[24] This lacuna can be fixed if we relax the notion of truth so that for functions f which do not have an inverse, the statement "$f = z$" ("f is the zero function") will be true. For example, we can define a probability measure m on S that measures the "size" (or the probability) of subsets of S. This will allow us to say, for example, that the probability that $f = z$ is true is equal to the size of the subset of s where $f = z$. Formally, $P_{f=z} = m(\{s \in S | f(s) = z(s)\})$, where $P_{f=z}$ is the probability that $f = z$. Using Boolean methods, one can extend this truth value measure to more complex sentences about R^S, where truth is now defined as measure 1 (so the measure of the set of exceptions is 0). We now denote with R^S/m the set R^S equipped with the new probabilistic notion of truth, determined by the measure m. It can be shown that R^S/m is always an ordered field.

We now take a set T whose cardinality is strictly bigger than the cardinality of reals. An example is the power set of the reals, whose cardinality is 2^c.[25] If I is the real unit interval, we define S to be the set $I^T = \{f : T \rightarrow I\}$. The members (points) on S are functions from T to I. The measure m on S is a natural extension of the Lebesgue measure on I. Our new model for the real line is $\{R^S/m\}$ where $S = I^T$, i.e., $\{R^{I^T}/m\}$. This model satisfies the 11 axioms of a complete ordered field.

Cohen used this model to demonstrate the falsity of the CH, as the new elements in $\{R^{I^T}/m\}$ play the role of points on a real line segment and have cardinality larger than $2^{\aleph_0}$ but smaller than c ([CH67], p. 116) . We, however, shall use it to construct a new metric structure for spacetime, different than the standard one based on the real line segment, with precise predictions that deviate from the latter.

2.5.2 *Ross' suggestion*

The idea, expressed in a (so far completely ignored) paper by the American physicist Dennis K. Ross [Ros84], is the following: once the real line model is replaced with its probabilistic alternative, the spacetime metric distances become random

---

[23] Note that as in the case of defining non-Euclidean geometries within the standard Euclidean space, we allow ourselves to use the real line model in defining its alternatives. Thus the following construction should not be interpreted as a replacement of the standard real line model, but rather as a consistent alternative to it.

[24] The proof for this failure is the following ([Lyn71], p. 94). Suppose S is the two-point set $\{0, 1\}$. Then each F can be defined by two real numbers $f(0)$ and $f(1)$. The identity element $u \in R^S$ is $u(0) = u(1) = 1$; the zero element $z \in R^S$ is $z(0) = z(1) = 0$. But now the function $h \in R^S$, defined by $h(0) = 0$ and $h(1) = 1$ fails to satisfy the multiplicative inverse, as $h \neq z$ and h has no inverse.

[25] By Cantor's diagonal argument, $|T| > c$.

variables, with dramatic physical consequences. The probability measure m involves a fundamental constant of length which modifies the short distance behavior of spacetime, and limits the resolution thereof.

Ross' idea is to replace the real line segment with an alternative $\{R^S/m\}$, as did Cohen, where m, the probability measure on S, is defined as follows:

$$m(\{s \in S | \beta \leq \chi(s, x) \leq \alpha\}) = \int_{\beta}^{\alpha} \frac{1}{(2\pi)^{1/2}} \frac{1}{\sigma} e^{-(s-x)^2/2\sigma^2} ds. \qquad (2.3)$$

Here each member of the standard real line segment $x \in \mathbb{R}$ has a random variable $\chi(s, x)$ associated with it, where $s \in S$ and $S = \mathbb{R}$. The random variable $\chi(s, x)$ is thus a real-valued function defined on the sample space S, with x denoting its mean. For $x = 3$, for example, equation (2.3) gives the probability distribution for the random variable associated with $x = 3$ of the standard real line segment. This random variable has a mean of 3 and a variance of σ.

The measure m gives the size of the interval $[\beta, \alpha]$ on S and is normalized to 1 when the interval includes the entire space S. The elements of the new line segment are the random variables $\chi(s, x)$. The fundamental constant of length σ measures the size of the spread of the probability distribution: as $\sigma \to 0$ the integrand in equation (2.3) becomes a delta distribution, and χ reduces back to its mean x. In this limit the new model reduces back to the standard real line segment.[26]

Ross points out that the above model which involves a normal (uniform) distribution and a constant σ, presumably of the order of the Planck length ($\sqrt{\hbar G/c^3} = 1.6 \times 10^{-33}$ cm) was chosen for its simplicity, but also because it suggests a physical picture of a short-distance foam-like spacetime structure where each point is treated equivalently, similar to some physical models arising from quantum effects in general relativity.[27] Lacking a guiding coordinating principle that picks up a specific R^S/m-type real line model,[28] what follows should only be considered as a heuristic plausible proposal.[29]

Importing this alternative model of the line segment to physics, one can define the actual length of path on a Riemannian manifold whose average length (in local

---

[26] The constant σ is the *same* for every random variable associated with every point in $\mathbb{R}$. However, even if not *all* functions from S to R are considered here, the model still satisfies the 11 axioms of a complete ordered field, see [Ros84], pp. 1211–1213 and p. 1218.

[27] The foam-like structure here results from the smearing of the metric distance, and not from topological changes that are characteristic of the models arising from quantum effects.

[28] Such coordinating principles, namely, the light principle and the equivalence principle, exist in the special and the general theories of relativity, respectively.

[29] It is noteworthy that Eddington, in his otherwise notorious *Fundamental Theory* ([Edd46], pp. 1–9) arrives at a similar Gaussian distribution for his definition of the scale of minimal observable length. Eddington's considerations, based as they are on the uncertainty principle, are strictly operational, and are similar to those appearing in modern non-commuting geometries. More on this issue in Chapters 4, 5 and 7.

coordinates) is

$$S = \int_{P_0}^{P_1} \left[g_{\alpha\beta} \frac{d\times^\alpha}{dP} \frac{d\times^\beta}{dP} \right]^{1/2} dP, \tag{2.4}$$

(where $g_{\alpha\beta}$ are the metric tensor components, and $\times^\alpha(P)$ is a piece of curve with $P_0 \leq P \leq P_1$), as the random variable ξ, where the probability that ξ will lie between the values α and β is given by

$$P_\xi = \int_\alpha^\beta \frac{1}{(2\pi)^{1/2}} \frac{1}{\sigma} e^{-(\xi - S)^2/2\sigma^2} d\xi. \tag{2.5}$$

The above mapping satisfies the conditions for a distance function (when the latter are interpreted probabilistically, see [Ros84], pp. 1214–1215).

Note that while ξ may take any value, only values of ξ within several σ of S are probable. As an example, take the propagation of light on "null" geodesics, whose nominal, or average, distance is 0. The actual length of the null geodesic is then the random variable ξ with probability

$$\int_\alpha^\beta \frac{1}{(2\pi)^{1/2}} \frac{1}{\sigma} e^{-(\xi)^2/2\sigma^2} d\xi \tag{2.6}$$

of being between α and β. The mean value of ξ is

$$\int_{-\infty}^\infty \xi \frac{1}{(2\pi)^{1/2}} \frac{1}{\sigma} e^{-(\xi)^2/2\sigma^2} = 0 \tag{2.7}$$

as we would expect. Physically, this means that the light cone is "smeared out" in a probabilistic sense by several orders of σ, and that we have lost strict relativistic causality on length scales of the order σ, while preserving it "on the average." Failure of relativistic causality is germane to many models of discrete spacetime, and has important consequences – theoretically as well as experimentally.[30]

2.5.3 Methodology

There exists an important methodological difference between the way discreteness enters into spatial considerations in the former cases of topology and geometry (Sections 2.2, 2.3, and 2.4), and the way it is considered here. While in all cases both (a) the structure of space and (b) its mapping onto the real line segment, are

---

[30] Ross suggests imposing experimental constraints on σ that stem from this feature, and even proposes a possible set-up that involves the Mössbauer effect [Eyg65, VF98] to calculate them. His rough estimation is of the order of $\sigma < 10^{-24}$ cm. The Mössbauer effect already appeared as a possible experimental set-up to probe the fundamental length scale (via the loss of relativistic causal order it generates) almost three decades earlier in the work of Alden Mead [Mea66]. Ross (private communication) had no knowledge of those earlier ideas.

part and parcel of the analysis and the argumentation, the order in which these two issues are considered is reversed.

The point is the following. Zeno's arguments, as well as the argument against collision and the tile argument, all start from a specific spatial structure and use the appropriate mapping to illustrate their respective claims. That this methodology is reversed in Ross' proposal signifies an operational, or epistemological, twist in the way spatial discreteness is conceived: one is no longer making claims about ontology and the structure of space. Rather, the claims are now restricted to our (in)ability to resolve distances smaller than a certain minimal length. The discreteness that arises in this sense is thus not necessarily a feature inherent to the structure of space; it could also be a feature inherent to physical measurements. This epistemological context, as we shall see, is precisely the one in which the notion of fundamental length entered modern physics in the 1930s.

2.6 The physical Church–Turing thesis

2.6.1 Mathematical logic

The mathematical model for a universal computer was defined long before the invention of digital computers and is called the Turing machine. That this model, which we now call an algorithm, captures the concept of computability in its entirety is the essence of the Church–Turing thesis (CTT henceforth), according to which every effectively calculable function can be computed using a Turing machine. Admittedly, no counterexample to this thesis (which is the result of convergent ideas of Turing, Gödel, Post, Kleene, and Church) has yet been found. But since it identifies the class of computable functions with the class of those functions which are computable using a Turing machine, this thesis involves both a precise mathematical notion and an informal and intuitive notion, and hence cannot be proved or disproved. Simple cardinality considerations show, however, that not all functions are Turing-computable (the set of all Turing machines is countable, while the set of all functions from the natural numbers to the natural numbers is not), and the discovery of this fact came as a complete surprise in the 1930s [Dav58].

The Turing machine model is discrete: it deals with sets of bit strings, or equivalently functions on the natural numbers $f : \mathbb{N} \to \mathbb{N}$. Moreover, every computation is a finite sequence of discrete computational steps, each requiring a finite amount of time. However, the model does permit potential infinity in its input and output size (the machine's tape is finite but also unbounded). This allows one to define [Tur37] the notion of a single computable real number: a real number is computable if its decimal expansion can be computed in the discrete sense (i.e., as an output of some Turing machine).

To deal with continuous functions, a notion of computability over the reals is required. There are now two main approaches to modeling computation with real number inputs. The first approach, usually called the bit-model [BC06], reflects the above mentioned fact that computers can store only finite approximations to real numbers. Roughly speaking, a real function f is computable in the bit-model if there is an algorithm which, given a good rational approximation to x, finds a good rational approximation to $f(x)$. The second approach is the algebraic approach [BCS97], which assumes that real numbers can be represented exactly and each arithmetic operation can be performed exactly in one step. The algebraic approach applies naturally to arbitrary rings and fields, although for modeling scientific computation the underlying structure is usually $\mathbb{R}$ or $\mathbb{C}$.

A further extension of computability over the reals was suggested by Moore [Moo96], which allows the computation of continuous functions to be modeled as a continuous process in time, rather than as a sequence of discrete steps. The class of computable functions under this formalization turns out to be quite large, and includes many functions which are uncomputable in the traditional sense. Moore further stratifies this class of functions into a hierarchy, according to the number of uses of a zero-finding operator μ, an operator that finds the "smallest" y such that $f(\vec{x}, y) = 0$. At the lowest level are continuous functions that are differentially algebraic, and computable by a general purpose analog computer. At higher levels are increasingly discontinuous and complex functions which are considered uncomputable in the discrete model.

As in the other domains we have inquired into before, here also the mathematical notion of computability seems to handle both the continuous and the discrete without distinguishing between them. And as in the other domains we have inquired into before, the issue at stake is not the consistency of the mathematical representation, but its applicability to the physical world. Although the original CTT was conceived in the domain of mathematical logic, physicists often interpret it as saying something about the scope and limitations of physical computing machines.

2.6.2 Physics

This physical version of the Church–Turing thesis (PCTT) says that any physical system can be simulated (to any degree of approximation) by a universal Turing machine [Wol85], and that complexity bounds on Turing machine simulations have physical significance. For example, if the computation of the minimum energy of some system of n particles requires at least an exponentially increasing number of steps in n, then the actual relaxation of this system to its minimum energy state will also take an exponential time. An even stronger version of the PCTT says that a probabilistic Turing machine can simulate any reasonable physical device at polynomial cost [Pit07].[31] Further variants of this thesis can be found in [Cop96].

In order for the PCTT to make sense we have to relate the space and time parameters of physics to their computational counterparts: memory capacity and number of computation steps, respectively.[32] There are various ways to do that, leading to different formulations of the thesis [Pit90]. For example, one can encode the set of instructions of a universal Turing machine and the state of its infinite tape in the binary development of the position coordinates of a single particle. Consequently, one can physically realize a universal Turing machine as a billiard ball with hyperbolic mirrors [Moo90, Pit96].[33]

Do physical processes exist which contradict the PCTT? There are at least two counterexamples to this thesis that purport to show that the notion of recursion, or Turing-computability, is not a natural physical property that must hold *a priori* (see [PeR81, Pit90, Hog94]). However, the physical systems involved in these coun-terexamples – a specific initial condition for the wave equation in three dimensions and an exotic solution to Einstein's field equations, respectively – are somewhat contrived, and, consequently, most physicists believe that the PCTT does hold in our world:

Does the physical Church–Turing thesis, that the physical world is computable, still hold? In a perfect, classical world where the μ operator [the zero-finding operator] can be implemented, no. But in a world with noise, quantum effects, finite accuracy, and lim-ited resources, even $\delta(x)$ is not physically realizable: how can we tell precisely whether $x = 0$ or not? If x is a velocity, we need to wait an infinite time to see if it moves; if x is a probability, we need an infinite number of ensembles to see if it happens; if x is a position, we need light of infinite frequency to locate it; if x is $T - T_c$ where T_c is a critical temperature, we need an infinite number of particles for the thermodynamic limit to be meaningful. So in the world we live in, only the lowest level of the μ-hierarchy seems to be realizable, and the physical Church–Turing thesis seems safe.

([Moo96], p. 43)

2.6.3 Existence (and sometimes uniqueness) of solutions

This strong belief in the applicability of the PCTT in the actual physical world is also manifest in the methodological tendency to constrain all of our current dynamical theories by the requirement of existence of solutions. Most working

---

[31] This stronger version of the PCTT is challenged by quantum computing. See [Hag06] for an accessible review.

[32] It should be stressed that there is no relation between the original CTT and its physical version [SP03], and while the former concerns the concept of computation that is relevant to logic (since it is strongly tied to the notion of proof which requires validation), it does not analytically entail that all computations should be subject to validation.

[33] For the most intuitive connection between abstract Turing machines and physical devices see the pioneering work of Gandy [Gan80], simplified later by Sieg and Byrnes [SB99].

physicists tend, as a matter of fact, to disregard naked singularities, closed time-like curves, non-globally hyperbolic spacetime models, ill-posed initial value problems, divergences, and the like, adhering to the constraint of existence of solutions to the dynamical equations.[34] This tendency is prevalent notwithstanding the logical possibility of singularities (that signify actual infinity of some physical magnitude) or non-recursive dynamical trajectories (that imply the physical realization of an infinite number of computational, i.e., dynamical, steps in a finite time), which means that, rather than signifying physical possibility, these cases delineate the point where physics actually breaks down.[35]

That the existence of solutions is a crucial feature here is also evident from the philosophical literature on Zeno's paradox of Achilles and the Tortoise. Commenting on this paradox, philosophers have argued [Bla50, Tho54] that the notion of a supertask [Lar09], or the completion of an infinite number of steps in a finite time, is incoherent, as it involves a final state when *ex hypothesi* there can be none. Contradiction only follows, however, if one assumes that every state must be logically dependent on its predecessors. This assumption amounts to requiring existence of solutions; it disallows a state which is undefined, or which is logically independent of any of its predecessors [Ben62] – think of a function f whose domain is $[0, 1)$. To ask what is the state of a physical system at the end of a supertask is to ask what is the value of $f(1)$. And yet, while supertasks are logically possible, no Turing machine can perform them as, by definition, every step of a computation is dependent on an earlier step. Probabilistic Turing machines do not violate this existence condition; they simply drop uniqueness.

Finally, note that the two properties, existence of solutions and spatial discreteness, need not contradict each other. On the contrary, in a world in which the former holds, no realization of a physical magnitude with a real number value is allowed (as, under the PCTT, the physical processes that are involved in this realization are uncomputable). The admissible computable functions (i.e., the allowed dynamical evolutions) are thus functions from the (bounded) set of rationals to the (bounded) set of the rationals (where the bounds depend on the size of the system at hand).

---

[34] If one insists that every dynamical evolution uniquely takes a physical state to one and only one physical state, one has what is sometimes called deterministic causation. A version of this which drops uniqueness but keeps existence is sometimes called probabilistic causation.

[35] It is interesting that classical Newtonian mechanics, while considered by many to be the paradigm for determinism, is actually much more hospitable to singularities (collision and non-collision alike) than quantum mechanics, and requires additional structure to rule these out, see [Ear86] and [Ear04]. Note also that from a strictly dynamical perspective, quantum *dynamics* is fully deterministic: Schrödinger's equation takes any quantum state to one and only one quantum state. The question of what this state stands for is a matter of ongoing debate, but the notorious projection postulate still defies precise physical formalization, and the indeterministic theories that do formalize it, e.g., [GRW86], are not part of standard physics. This is one of the facets of the quantum measurement problem.

That these functions are undefined for irrational values is, of course, not a genuine problem, as they are not required to be so defined.

2.7 Pure versus applied mathematics

Zeno's paradox of extension, as well as the argument against collision and the arguments for or against the possibility of supertasks, can be seen as challenging the consistency of certain mathematical descriptions of space, and in particular of the mathematical representation of the line segment. The lesson there, as we have seen, is that once we allow input from modern mathematics and modern physics, and use this input to refine our pre-theoretical definitions, both a description of the line segment which is continuous (whether countable or uncountable), and a description of the latter which is finite and discrete, are perfectly consistent.

The issue at stake, then, is not the consistency of these structures, but rather their applicability to the physical world: which mathematical description, the discrete or the continuous, is an adequate description of physical reality? And if both are adequate descriptions, then which one is more fundamental?

Weyl's tile argument, as well as Grünbaum's criticism of the method of abstract extension, aim to establish that only one of the two structures, namely the continuous, is an adequate representation of physical reality, and that there is no way even to approximate the continuum with the discrete. Weyl and Grünbaum argue, in other words, that we have to make a choice, because the two structures cannot co-represent, each at its appropriate level of applicability, our physical world.

The remarkable applicability of the continuum in describing macroscopic physical space cannot be taken for granted as evidence for its apparent uniqueness; as in the case of non-Euclidean geometries, the fundamental nature of physical space may well be different, and clearly need not be decided by mere *a priori* arguments. As we have seen, there are ways to rebut the tile argument with discrete microstructures that can approximate the continuous macrostructure. Moreover, there are fine grained probabilistic alternatives to the real line segment which involve a minimal length, or a finite spatial resolution, that reduce to the standard real line segment when coarse grained, and that may carry an empirical signature.

What does this short excursion into the realms of pure and applied mathematics teach us? First and foremost, we learn that no mathematical or geometrical structure of physical space can be dictated by logic. Current physics widely employs the notion of the continuum, but ultimate physical reality may be fundamentally different. Can developments in physics help us tell the difference? While the question of applicability may well be empirical, it is also contingent on resolution capabilities. Deferring the discussion on phenomenology to the end of the book (see Chapter 8), on the theoretical front we now know that the important question is not whether alternative discrete models of space are logically possible, but rather which discrete model yields new predictions, while satisfying best one of the following two requirements: (1) reproducing the same results as the current continuum model (e.g., the Pythagorean theorem in macroscopic space, or the results obtained by standard integral and differential equations), or (2) approximating these results of contemporary theory to a high degree of accuracy.

3

Arguments from philosophy

3.1 Outline

Spatial discreteness – interpreted either as an ontological commitment, or as an epistemological constraint on measurement resolution – cannot be ruled out as a matter of logic; neither can attempts at approximating the continuous with the discrete. This much we have learned from the previous chapter. But the metaphysics and the epistemology that may underlie, or motivate, spatial discreteness both carry price tags, and philosophers, who are in the business of exposing these tags, are quick to point them out.

In this chapter we shall inquire about possible philosophical reasons to reject spatial discreteness, that arise from worries about the consequences of holding on to a finitist metaphysics or to an operationalist epistemology that is consistent with this view.

One way to define finitism is by negation; it is the opposite of Platonism. Platonism with regard to mathematics is the metaphysical view that there are abstract mathematical objects whose existence is independent of us and our language, thought, and practices. The continuum, or actual infinity, according to the mathematical Platonist, is thus a viable possibility, irrespective of our inability to realize, observe, or measure such a magnitude. Mathematical finitists, in contrast, reject actual infinity exactly because of these inabilities. The continuum for the finitist is nothing but a mere fiction, and has no representation in the actual world.

A more positive definition of mathematical finitism categorizes it as a branch of intuitionism. Intuitionists see mathematics as a free creation of the human mind. According to them an object exists if and only if it can be (mentally) constructed. Intuitionists distinguish themselves from Platonists by the strict interpretation of the phrase "there exists" as "we can construct." They would thus replace actual infinity with potential infinity: we can get as close as we want to actual infinity, say, as we do with the modern $\epsilon - \delta$ concept of *limit*, but we can never attain it.

Finitists, however, while accepting the intuitionist premise about constructivism, would reject potential infinity: they would allow only bounded magnitudes, and so magnitudes that can increase arbitrarily would be considered unacceptable.

A central thesis of this monograph is that one can map these three approaches in mathematics, namely, the Platonist, the intuitionist, and the finitist, onto the practice of theoretical physics in the struggle with the notion of fundamental length. The next chapter will attempt to substantiate this thesis with historical evidence, but first we shall look at philosophical arguments that are marshaled against finitism, either as a metaphysical view about the (in)existence of actual infinity, or as a methodological thesis about the precedence of epistemology over ontology.

One metaphysical view consistent with finitism that may underlie spatial discreteness is computationalism. The basic building blocks of this so called "digital metaphysics" are cellular automata which inhabit (or give rise to) a discrete space-time. Digital metaphysics is hostile to actual infinities – consistent as they are with the abstract notion of computability; real numbers, which require infinite precision, are nevertheless excluded if one insists on (1) closure, i.e., that any input is also an output of some finite computational process, and (2) causation, deterministic or otherwise (see Section 2.6.3). Insisting on (1) and (2) is tantamount to rejecting Platonism and endorsing finitism. Some argue that the latter view carries with it a certain price: it does not allow one to make a distinction, highly regarded among philosophers, between metaphysics (the study of what *is*) and epistemology (the study of what can be *known*).

That such a distinction even exists is, of course, something that a finitist will gladly reject, but if one translates this challenge into a challenge about the interpretation of one of our best theories, namely, quantum mechanics, it becomes more pressing. The problem is that, for a finitist, there seems to be no difference between the uncertainty that exists, for example, in deterministic chaos, and the uncertainty that exists in the subatomic world and is captured by the formalism of quantum mechanics. This blurring of classical and quantum uncertainties stands in conflict, however, with the famous "no-hidden-variables" proofs, and the violations of Bell's inequalities, and is thus considered too high a price to pay. As we shall see here, the finitist can meet this challenge by putting forward a new notion of deterministic objective chance that would make the difference between classical and quantum uncertainties quantitative rather than qualitative.

The finitist faces a similar challenge in matters epistemological. Here the worry is the following. By endorsing the kind of operationalism and empiricism that motivate the finitist view, one subordinates what is to what one can know (or measure). In so doing, one risks making the original question about spatial discreteness completely moot from a scientific perspective, as it becomes purely metaphysical, or a matter of mere convention, that cannot be decided unequivocally, and transcends any experiment. To meet this challenge, a short detour to the philosophy of geometry may prove instructive, where an operationalist argument made by Einstein about the primacy of geometrical notions is shown to be independent of conventionalism or transcendentalism. In other words, the latter two do not necessarily follow from the operationalism that motivates finitism.

To strengthen this point further, we address the issue from yet another angle, making an analogy with another famous distinction in physics, namely the distinction between determinism and indeterminism, that is commonly deemed *transcendental* or metaphysical. Here, again, the finitist can pull herself out of the muddle, by showing how the alleged underdetermination between deterministic and indeterministic dynamics is broken when one accepts – as the finitist does – that the available physical resources, required to model an indeterministic dynamical flow with a deterministic one, are finite and admit an upper bound.

The upshot, as one might have guessed, is that there are no convincing philosophical arguments against spatial discreteness and the metaphysics or the epistemology that may underlie it. The purpose of this chapter is thus to convince the reader that whatever philosophical price tag finitism may carry, modern physics can afford it.

3.2 Metaphysical motivations

3.2.1 Digital metaphysics

Metaphysics is the study of ultimate reality, and as such it comes in many flavors. Physicalists believe that metaphysical reality and physical reality are identical, and that metaphysics reduces to physics. For dualists, on the other hand, ultimate reality consists of both material and immaterial substances – physics only encompassing the former. One of the possible metaphysical theses that underlie spatial discreteness, however, is different. It differs from physicalism in that it regards metaphysics as the study of the foundations of physics, and so it rejects the above reduction; it differs from dualism in that the duality it espouses is not a duality of substances, but a duality of function. Ultimate reality, according to this view, is computational; it is like hardware, on which physical reality runs as software [Fre90, Ste98].

The basic building blocks for this hardware are cellular automata:

Discrete in space. They [cellular automata] consist of a discrete grid of spatial cells or sites. Discrete in time. The value of each cell is updated in a sequence of discrete time steps. Discrete states. Each cell has a finite number of possible values. Homogeneous. All cells are identical, and are arranged in a regular array. Synchronous updating. All cell values are updated in synchrony, each depending on the previous values of neighboring cells. Deterministic rule. Each cell value is updated according to a fixed, deterministic rule. Spatially local rule. The rule at each site depends only on the values of a local neighborhood of sites around it. Temporally local rule. The rule for the new value of a site depends only on values for a fixed number of preceding steps (usually just one step).

([Wol86], p. 1)

The thesis that ultimate reality is computational is not new. Leibniz already envisioned the world as an automaton in his "Monadology," Charles Babbage thought natural laws were like programs run by his analytical engine, and recently, under the PCTT, dynamical evolutions are regarded as computations (see Section 2.6). This view of physical processes as computational processes rejects actual infinities. To quote Ed Fredkin, one of its modern supporters, it sees the physical world as "a large but finite system; finite in the amount of information in the total volume of space-time, and finite in the total volume of spacetime" ([Fre90], p. 255). Consequently, space and time are only finitely extended and divisible, all physical quantities are discrete, and all physical process are only finitely complex. Infinitely complex entities such as the Mandelbrot set [BC06] are rejected as physically unrealizable.

The thesis maps directly onto finitism in the philosophy of mathematics: while the ultimate goals of the mathematician are completeness and closure, the means for finding solutions to applied problems are always limited, dependent as they are on computationally feasible tasks. Finitism in mathematics is therefore analogous to applying digital metaphysics to physics.

3.2.2 Platonism versus finitism

The traditional, Platonist, approach, which accepts Cantorian actual infinities, employs a mathematical machinery that has much more structure than is required to model physics, and a lot of effort is spent in disabling or reinterpreting these redundancies, so that the modeling can be done in spite of them. Once more let us take Maxwell's equations as an example: the state space for these equations is of cardinality $(\mathbb{R}^6)^{\mathbb{R}^3}$. To deal with this uncountable set of states, we concentrate on systems having very special properties – for example, continuity, uniformity, locality, linearity, or reversibility – Maxwell's equations happen to have all of these properties at once [Tof84]. These features allow us to disregard most of the infinities. Continuity and linearity, for example, mean that a small perturbation in the system's initial state leads to a correspondingly small perturbation in its final state, so that we do not have to worry about capturing its state with infinite precision, as any error in this description would be bounded.

On the other hand, in the finitist approach the theoretical mathematical apparatus in which we "write" our models is essentially isomorphic to the concrete computational apparatus in which we "run" them. If one starts from this finitist approach, the only infinity that one may still want to incorporate in a physical theory is potential infinity, for example, one in which the size of a system is always finite but may grow arbitrarily, commonly defined by means of the $\epsilon - \delta$ limit concept. Note, however, that the natural topology in which to take this limit in this approach is that of the Cantor set, rather than that of the real numbers:

> The customary picture which represents the state of a system as a point tracing an orbit in phase space is somewhat misleading: in general, we cannot handle in a finitary way the exact position of the point itself (which encodes an infinite amount of information); on the other hand, we can project the point on a finite subset of axes, and we can enlarge this subset as need requires. We shouldn't try to (and, at any rate, we can't) take in the whole picture!
>
> *([Tof84], p. 126)*

The finitist challenges the very idea of "independent existence" in mathematics, because unless an existence claim is attached to a definite procedure to verify it, it is not clear what its reference is. Consider Goldbach's conjecture: "Every even number greater than 2 can be represented as the sum of two primes." This conjecture is still undecided, and so, according to the finitist, as long as there is no effective procedure to prove or disprove it, it has no truth value.

The finitists believe that the Platonic position – that statements about numbers are true or false independently of our knowledge, as if the totality of natural numbers were objects that exist and possess properties independently of human minds and of human knowledge – is ultimately indefensible. Truth for a finitist is thus identifiable with provability.[1]

3.2.3 Determinism versus predictability

To emphasize the difference, Pitowsky makes an analogy between the rival camps in the philosophy of mathematics and the opposing views about determinism in physics [Pit96]. On the Platonist view, determinism is an ontological notion – it has to do with the existence of a set of unique future states given the present state of affairs. These states have properties exactly in the same way that natural numbers have properties, regardless of anyone's knowledge. In contrast, on the finitist view, determinism is an epistemic notion; it has to do with the predictability of a future state from the given current state of affairs. Laplace's famous demon has the ability to predict with certainty; we humans are finite creatures, with limited resources, hence are bound to live with uncertainty in our predictions.

The finitist perspective, however, does come with a price. Given the PCTT (the thesis that physical processes are algorithms), uncomputable processes are excluded from physics, and oracles do not exist in the physical world. This restriction means

---

[1] Note that finitism is stronger than mere intuitionism, where it is sufficient that an algorithm, albeit unknown to us, exists that can, in principle, decide a proposition. Finitism disputes even this contention, and only admits feasible proofs, e.g., such that their time complexity is bounded by a polynomial.

that many questions – those questions whose answer requires quantification over unbounded time for which no algorithm that can decide them exists – are physically meaningless. These questions may seem completely legitimate for the Platonist, for example, "is the solar system stable or will it collapse one day?," or "is this physical motion periodic?," but are in effect totally unscientific from the finitist perspective, as there is no feasible way to answer them.

Another common accusation against finitists is that they blur the difference between ontology and epistemology by equating determinism and predictability [Ear86]. Clearly there are perfectly deterministic processes which cannot be predicted, for example, processes whose dynamics are sensitive to initial conditions and are characterized as chaotic. This means that lack of predictability (as an epistemic notion) does not entail indeterminism (as an ontological notion).

In defense, the finitist could say that this accusation begs the question: the alleged difference is exactly what she denies, since the only way to make physical sense of an ontological claim is with an epistemological one, namely, with an experiment. But while this defense is perfectly legitimate, it leads to another, more interesting worry.

3.2.4 Deterministic objective chance

The worry is the following. On the finitist view, classical deterministic chaotic systems and quantum systems are both unpredictable hence indeterministic, but in that case, what is so revolutionary about quantum mechanics? Is it just a matter of degree? Is it merely the fact that the unpredictability is more prevalent in the subatomic world? The received view, for example, [Rei42], is that even if Laplace's demon were endowed with perfectly ideal measuring devices and all the computation powers of an oracle, he would still not be able to predict the future of a quantum system. Support for this view comes from the Copenhagen interpretation of the uncertainty principle, from the famous "no-hidden-variables" proofs [KS67], and from the confirmed violations of Bell's inequalities [A+82]. Consequently, by failing to distinguish (un)predictability from (in)determinism, finitists seem to blur the difference between classical and quantum mechanics ([Pit96], p. 176), as they cannot distinguish between apparent, subjective randomness, and real, objective chance.

Finitists can only "bite-the-bullet" here [Mey99], denying as they do any difference between appearance and reality, but the situation is not as bad as it seems. Agreed, if the PCTT holds, all dynamical evolutions are deterministic, and indeterminism can only mean lack of predictability. But one can still interpret "lack of predictability" as an objective feature of the world, its deterministic dynamics notwithstanding. One example is the view known as *typicality* (this approach is advocated both in classical statistical mechanics and in Bohmian mechanics [Mau07]). Here, lack of predictability arises from a genuine, objective, feature of the world, namely, the initial distribution of particles.[2] True enough, Bohmian mechanics and its objective interpretation of deterministic chance have their problems,[3] but (1) these problems do not arise in any way from the interpretation of probabilities as objective, and (2) there is no argument that I know of that shows there can be no alternative to such an objective interpretation which sidesteps these problems.[4]

The finitist traces the origin of the probabilities that arise both in classical and in quantum mechanics to objective finite resolution, which is, as a matter of fact, the state of affairs for any physical measurement we have done, and will ever do. In this sense the finitist rejects any qualitative, metaphysical, difference between the two. The only difference is that while in classical mechanics infinite resolution is possible *in principle*, in quantum mechanics it is not. And yet, that there is no qualitative metaphysical difference does not mean that there is no difference at all. For example, the finitist can still point at a quantitative difference in *measure* between classical and quantum probabilities, which is captured by quantum non-commutativity, manifest in the violations of Bell's inequalities, and is believed by many to be the source behind the putative quantum speed-up of computation (see, e.g., Pitowsky [Pit07]): if one allows only local interactions, the classical (Boolean) probability theory gives different predictions than the quantum (non-Boolean) probability theory for the same Bell-type experiment. This (quantitative) difference is manifest in the finitist, objective, alternative, which, under the assumption of the locality of interactions, reproduces the Born rule for all experimental set-ups we have done so far [HS11].[5] Metaphysical issues or questions, for example, of whether a system has a definite state when not being measured, have nothing to do with it.

---

[2] In Bohmian mechanics this initial distribution is postulated in such a way so as to reproduce the Born rule, see [GDZ92].

[3] In a nutshell, there are two main issues here. First, the notion of typicality, upon which this view is based, is unjustifiable, in the sense that its justification requires solving the problem of induction [Pit85]. Second, while empirically indistinguishable from orthodox quantum theory, Bohm's theory is genuinely non-local, and as such resists relativistic generalizations, see [Alb92], Chapter 7.

[4] For a possible alternative see [HS11] where a finitist, dynamical, and objective notion of chance is defined and defended, which is based on the notion of physical computational complexity.

[5] In a nutshell, the idea here is to view classical and quantum probabilities as transition probabilities between any two physical states. Given a fixed amount of physical resources (energy, space, time), one assigns probability to such a transition according to the "number" of possible dynamical evolutions that can realize it, weighing these inverse proportionally to their respective physical computational time-complexity. The said violations of Bell-type inequalities indicate, or so the story goes, that quantum dynamics "exploit" more efficiently the difference in physical resources that "separate" any two physical states. For a fixed amount of physical resources, and a certain size of the physical system, there will be a quantitative difference in probability measure (defined as the "size" of the set of possible dynamical evolutions that can realize the transition between the two states) that can distinguish the classical case from the quantum one. Note that, contrary to Bohmian mechanics (which always reproduces the Born rule, and so is empirically indistinguishable from quantum theory), this model is *in principle* distinguishable from quantum theory (although it agrees with it on all experiments we have ever done): since the probability measure in this model depends on the system's size, deviations from Born's rule would only arise in situations that involve measurement apparatuses made of very few particles.

That this finitist perspective can motivate an alternative interpretation of quantum theory – a deterministic interpretation which at the same time is agnostic about metaphysics – is also evident from the pivotal role it assigns to measurability and to measurement outcomes. This role is also manifest in the operationalism of orthodox non-relativistic quantum mechanics, where measurement outcomes are taken to be primitive notions [Pei91], and the generalization of the theory to the relativistic regime requires a thorough analysis of the measurability of field quantities [BR33, BR50]. This analysis, carried out by Bohr and Rosenfeld in the early 1930s, establishes a framework in which measurable quantities are always finite, and measurement is always done with finite resolution. We shall develop this finitist reasoning further in Sections 4.3.1 and 5.6.

3.3 Epistemology and the primacy of "length"

3.3.1 Geometry and experience

In his celebrated paper "Geometrie und Erfahrung" (Geometry and Experience) [Ein21], first presented as a lecture to the Prussian Academy of Sciences at Berlin in January 1921, Einstein made a clear distinction between pure and applied geometry. The former, he said, is a mere axiomatic system that acquires its certainty from its "formal-logical" character, whose basic building blocks, for example, primitive terms such as "point," "line," "congruence," and so on, do not refer to objects or concepts "in the world," but rather derive their meaning from the axioms. These axioms serve as "implicit definitions" of the primitive terms, and all the theorems of mathematical geometry follow purely logically therefrom. The latter, in contrast, arises when one gives some interpretation of the primitive terms via real objects of experience, coordinating the "empty conceptual schemata" of pure geometry with real physical (rigid) bodies.

This coordination transforms geometry into an empirical science, but it does come with a price:

In so far as the propositions of mathematics refer to reality they are not certain; and in so far as they are certain they do not refer to reality.

([Ein21], p. 233)

So the purity and certainty of mathematical geometry as a logical, deductive, system is permanently lost when we interpret its basic building blocks by coordinating them with physical objects. What we gain, however, is an empirical science that can yield falsifiable predictions.

The view Einstein espouses was taken by logical-empiricists to refute the Kantian conception of geometry as a synthetic *a priori* science (see, e.g., [Hem45]). Einstein was well aware of the neo-Kantian tradition, that flourished in the nineteenth century through the works of Felix Klein, Bernhard Riemann, Sophus Lie, and Herman von Helmholtz (see [Fri02] for a comprehensive account). According to this tradition, pure geometry was not just an empty logical structure, but a description of the most general and abstract features of a human's perception of space. The most important results of this tradition are the so called Helmholtz–Lie theorems (see [Ste77], pp. 21–25, especially fn. 29): starting from basic kinematic conditions of "free mobility" that allow continuous motion of rigid bodies, one can derive the Pythagorean form of the line element, and thereby fix the geometry of space (up to a scale factor) to be of constant curvature – Euclidean, hyperbolic, or elliptic. That the former holds in the physical world would mean that measurements made with rigid bodies would approximate its laws to a high degree of accuracy.

This nineteenth century neo-Kantian view of geometry should also be contrasted with the so called "conventionalist" approach, best articulated by Henri Poincaré [Poi02]. While agreeing with the neo-Kantians about the results (and the conceptual background) of the Helmholtz–Lie theorems, Poincaré disagrees that the actual curvature of space could be "read off" with measurements of rigid bodies, as this presupposes knowledge about physical rigidity, i.e., knowledge about forces acting on the material constituents of bodies, and this knowledge requires geometry, if only to make spatial measurements that can confirm it. But if experience cannot pick up a geometry, we are left with a choice of convention:

> Geometry is not an experimental science; experience forms merely the occasion for our reflecting upon the geometrical ideas which pre-exist in us. . . . Our choice is therefore not imposed by experience. It is simply guided by experience. But it remains free; we choose this geometry rather than that geometry, not because it is more true, but because it is more convenient.
>
> ([Poi98], pp. 41–42)

The issue at stake here is the theoretical status of rigid bodies. Helmholtz takes rigid bodies to be primitive, and uses them to fix the geometry of physical space, while Poincaré sees them as material objects, which, as far as physics goes, are subject to further investigation with dynamical theories that describe their microconstituents. Yet these dynamical theories must be confirmed somehow with measurements which presuppose rigid bodies, hence the inevitability of geometry.

It is noteworthy that Einstein, after some deliberation, takes sides in his paper with Helmholtz: while admitting Poincaré has a point in that rigid rods are material objects and hence require further dynamical analysis, he nevertheless prefers to treat them as primitive entities. Michael Friedman ([Fri02], p. 207) argues that Einstein's insistence on the primitivity of rigid bodies is closely related to the process by which he arrived at the general theory of relativity from the special one, and as far as the history of general relativity is concerned, this may as well be true.

But one can agree with the historical remark, and still offer another interpretation of Einstein's choice that can clarify the deep epistemological issue raised by Poincaré and Helmholtz, to which Einstein was perhaps sensitive. This interpretation will also help us elucidate yet another putative price that finitism – motivated as it is by the epistemological thesis of operationalism – may carry.

3.3.2 Geometry versus dynamics

The status of geometry depends, as we have seen, on one's attitude towards rigid bodies. Poincaré demands that they should be subject to further dynamical analysis; Helmholtz and Einstein, for different reasons, take them to be primitive. Helmholtz and Einstein also derive different conclusions about the metric of physical space from this irreducibility – whether it is Euclidean or non-Euclidean – while Poincaré insists that the question has no real physical significance and is merely conventional. What makes this state of affairs interesting and relevant to our discussion on the price of a finitist epistemology, is that it allows us to distinguish operationalism – the thesis that sees measurable quantities as primary – from two other views about geometry, namely, *a priorism* and conventionalism.

At stake lies a question of epistemic precedence. That rigid bodies are primitive means that in a certain epistemic sense the notion of "length" is irreducible and, in particular, not reducible to anything dynamical; one must have this notion in place prior to identifying dynamical objects as geometrical, and prior to testing the empirical predictions of the dynamical theory. Viewed from this perspective, the argument Poincaré makes in support of his conventionalism – a conventionalism about the actual metric of physical space – is exactly an argument that can support the irreducibility of geometrical concepts. In fact, in other places Einstein makes the same argument when he defends the precedence of geometrical notions against attempts to derive them from any dynamical theory of matter – presumably more fundamental (we develop this issue further in Chapter 6).

The argument is the following. Take two spatial points A and B and consider the distance between them as measured (in some Lorentz reference frame F) by rods and clocks. Einstein ([Ein21], p. 237) calls this distance *Strecke*, which may be interpreted as "stretch," or "segment" (see Section 6.5.2). Now one can argue that the fact that in F the distance between A and B is given by $AB_{segment}$ cannot be derived from dynamical considerations. Indeed, it is a sort of fact that cannot be derived at all.

The reason is that in order to make the linkage between the dynamics and geometry there must be a prior interpretation associating certain magnitudes in the dynamical theory with primitive geometrical notions such as "segment." This association is not dictated by the dynamics, but is rather presupposed as a constraint before the dynamics is written down, in order to make contact between empirical predictions derived from the dynamics, and our experience of the outcomes of measurements carried out by means of rods and clocks.[6] Without a primitive notion of "segment" in the first place, one cannot make contact between the dynamics and experience, since, as a matter of fact, the only kinds of things we can measure are various sorts of segments.[7] In our theories of space and time (and of spacetime) we construct structural laws (e.g., equations of motion) in precisely such a way that they will satisfy this primitive (geometric) interpretation. Only when such an interpretation is given can we test the dynamical theory experimentally. In other words, the notion of "segment" is a pre-requisite for experimentally testing the dynamical theory from which one wishes to derive the notion of "segment" from the outset.

Note that this argument is not a general argument against reduction. For example, one can still attempt to reduce thermodynamics to the kinetic theory of gases, because, contrary to the geometrical concept of "segment," the concept of "temperature," say, need not be assumed *ab initio* in order for us to verify the predictions of the kinetic theory of gases in spacetime. Those predictions can be measured without any recourse to temperature, since we actually measure segments when we measure temperature. But this is not the case when one tries to derive geometrical notions; here one must assume these notions in advance. The argument thus exposes the epistemic incoherence latent only in those eliminative attempts to derive from a reducing theory concepts that appear as primitives in the reduced theory, when the empirical confirmation of the former depends on the latter.[8] What it shares with operationalism, as well as with the finitist approach and the positivist attitude manifest in, for example, Whitehead's method of extensive abstraction, is the premise about the theoretical primacy of measurable magnitudes.

Salmon ([Sal01], pp. 36–38) argues that an operationalism based on this premise is too narrow. The worry here is that if one accepts that concepts such as "segment"

---

[6] One should be careful here: the point is not that *this* or *that* piece of empirical data constrains our dynamical theories; rather, it is that the whole idea of science as an empirical endeavor – the fact that we use measurements to verify our dynamical theories, and these measurements rely on geometrical concepts – constrains those dynamical theories to designate *ab initio* some theoretical magnitude as "segment," hence to presuppose this geometrical notion as primitive.

[7] More generally, the operational view expressed here is that all measurements, even measurements of temperature, intensity, or what have you, are ultimately position measurements, and involve resolving distances, or wavelengths, using "measuring rods," (see also [Bel87], pp. 52–62). Note that this does not necessarily commit one to the stronger, ontological view, found in, e.g., Bohmian mechanics, according to which all physical properties supervene on position.

[8] A similar incoherence exists in the attempts of some quantum gravity theories to derive the notion of time from dynamical considerations (e.g., [Bar00]). For a criticism of such a "timeless" quantum gravity see [Hea02]. Chapters 6 and 7 develop this thesis on the primacy of geometry further.

or "distance" are primitive, and if one further subordinates their characteristic features to one's epistemology, one ends up dictating ontology from epistemology. To make an analogy between the nineteenth century debate (on non-Euclidean geometry) and the debate we are interested in here (about discrete versus continuous geometry), if one builds one's ontology from actual measurement results, as, for example, Whitehead demands, then this ontology must have the same features or properties of actual measurement results, namely, it must be finite and discrete.

But it is important to emphasize that neither conventionalism nor pessimistic neo-Kantian *a priorism* necessarily follows from this platitude, be it in the context of spatial curvature or in the context of spatial discreteness, and that it still makes sense to ask what is the exact state of affairs. Helmholtz might have thought that since the notion of "segment" is based upon the motion of a rigid body (via the notion of congruence and the Helmholtz–Lie theorems), either the basic spatial relations of one of the three constant curvature geometries must hold, or geometry simply breaks down. But that Helmholtz' pet geometry holds strictly in our world does not follow from any psychological fact about the genesis of our geometrical concepts, nor does it follow from the Helmhotz–Lie theorems. At most what follows is that the postulates of Euclidean geometry hold approximately in our world, and besides, one can still look for other evidence that can settle the question. Riemann, in his celebrated paper "On the Hypotheses which Lie at the Bases of Geometry" (mentioned but not quoted in Friedman [Fri02] in his comprehensive analysis of "Geometrie und Erfahrung"), is clearly aware of this point, when he remarks:

Now it seems that the empirical notions on which the metrical determinations of space are founded, the notion of a solid body and of a ray of light, cease to be valid for the infinitely small. We are therefore quite at liberty to suppose that the metric relations of space in the infinitely small do not conform to the hypotheses of geometry; and we ought in fact to suppose it, if we can thereby obtain a simpler explanation of phenomena.

([Rie73])

Empirical knowledge about spatial relations may thus come from any source relevant to fundamental physical theory. The operationalist argument about the primacy of "length," or "segment," as geometrical notions, rather than establishing *a priori* that space must be of a certain character, simply reminds us of the epistemic constraint that in any such theory we still need to designate some physical magnitude as "length" of a line segment in order to make empirical sense of the theory's predictions. This designation, however, does not entail that the question whether this line segment is discrete or continuous is purely metaphysical or conventional, in the sense that it lacks empirical content. Once again, there are no good reasons for saddling finitism (and the operationalist epistemology that motivates it) with such a heavy, unaffordable, price.

3.4 Is discreteness transcendental?

That a certain question is transcendental, or metaphysical, means that there is no experiment that can settle it. Many believe that the question whether or not our world is deterministic is such a metaphysical question. As we shall see, however, finitism may help decide this, at least to some extent, and so, in a certain sense, the question of the nature of space, whether it is discrete or not, turns out to be epistemically prior to the question of determinism.

One philosopher who inquired about problems whose solutions transcend experience is Kant. In his *Critique of Pure Reason*, Kant poses these problems as "antinomies" – pairs of contradicting claims, "thesis" and "antithesis." Both seem to be equally supported by *a priori* arguments; neither can be decided by "appearances," as their solution exists outside the boundaries of time and space. The third of these antinomies concerns free will. Roughly speaking, it contrasts determinism (every event has a cause) with its negation.

Kant's third antinomy has an analogous conundrum relevant to our discussion. Without delving into the intricacies of Kant's arguments for either of the contradicting theses, or for his own resolution of the third antinomy, we can ask a similar question: which dynamics, deterministic or indeterministic, governs the world, and will we ever be able to decide this issue?

3.4.1 Methodology makes a difference

Several arguments support the claim that the question whether nature is deterministic or not cannot be settled with an experiment, since any amount of data (or measurements) can support both alternatives equally well [Sup93]. The richness and complexity of deterministic systems, for example, suggest that any phenomenon can be accounted for as governed by deterministic dynamics by a suitable choice of system type and suitable parameters. Put differently, it is always possible, or so the story goes, to rule determinism *in*, as any non-deterministic dynamics can be simulated by a deterministic one by adding additional structure to the theoretical apparatus.

Consider collapse alternatives to non-relativistic quantum mechanics such as the GRW theory [GRW86]. These theories alter the deterministic evolution; they do so by adding a stochastic, a-causal, element to the dynamics in such a way that gives the same predictions as standard non-relativistic quantum mechanics in all the experiments done up until now, but *in principle* gives different predictions in specific experimental set-ups that have yet to be performed. Now suppose that, some time in the foreseeble future, experimental set-ups of the sort that presumably could distinguish between a GRW-like theory and standard non-relativistic quantum mechanics are performed, and their outcomes turn out to be in agreement with the GRW-like theory to a very good approximation. Nevertheless, it is often argued that standard non-relativistic quantum mechanics would remain intact for the following reason.

Let us suppose that an open system S is subject to perfect decoherence, that is, to interactions with some degrees of freedom in the environment E, such that the states of the environment become strictly orthogonal. Suppose further that we have no access whatsoever (as a matter of either physical fact or physical law) to these degrees of freedom. In this case, the GRW-like dynamics for the density operator of S would be indistinguishable from the dynamics of the *reduced* density operator of S obtained by evolving the composite quantum state of $S + E$ unitarily and tracing over the inaccessible degrees of freedom of E.

It turns out that this feature is mathematically quite general, because GRW-like dynamics for the density operator is a completely positive linear map [NC00, S+01].[9] From a physical point of view, this means that the GRW-like theory is empirically equivalent to a quantum mechanical theory with a unitary (and linear) dynamics of the quantum state defined on a larger Hilbert space. In other words, one can always introduce a new quantum mechanical ancilla field whose degrees of freedom are inaccessible, and "cook up" a unitary dynamics on the larger Hilbert space that would simulate the GRW-like dynamics on the reduced density operator. Therefore, experimental results that might seem to confirm (indeterministic) GRW-like dynamics could always be re-interpreted as standard (deterministic) Schrödinger-equation-like quantum mechanics on larger Hilbert spaces.

And yet this reinterpretation is a double-edged sword. There is little merit in claiming that a theory cannot be proven wrong, as this also means that, from a scientific perspective, it cannot be proven right. Even a die hard empiricist need not accept the myth of underdetermination of theory by evidence [Lau90]; that it is logically possible to rule *in* determinism need not entail that it is also methodologically correct to do so. And while it is true that it may be hard to formalize the rules of "good sense" under which the scientific enterprise takes place, many physicists will nevertheless agree that a methodology that keeps hiding its anomalies and explaining them away with additional mathematical structures is nothing but suspect (for more on this so called "ancilla argument" see [Hag07]).

---

[9] Such a map can be considered as a physically acceptable dynamical evolution as it transforms any physical state to another even in the presence of entanglement.

3.4.2 Asymmetry

The methodological asymmetry described above is a well-known way out from arguments of underdetermination. Here, however, I would like to make an even stronger claim. I shall argue that there is no need to enter into a discussion on methodology; from a finitist perspective, the mere logical possibility (of accounting for an indeterministic dynamics with a deterministic one) is simply misleading.

Here is how the argument goes. If indeterminism holds, then a random sequence should be seen as "truly random," and not a result of ignorance of the initial conditions. But the notion of randomness depends on the physical resources involved [Cha75, For83]. In the quantum case, these resources are encoded in the dimension of the Hilbert space (i.e., whether or not the ancilla field above is physically real). In the classical case, these resources are manifest in the partition of phase space, or, equivalently, in the level of coarse graining one uses:

Whether or not a deterministic system represents a Bernoulli process or an (n-stage) Markov process is relative to a partitioning of the state space. The very same set of trajectories or orbits may in one case yield itineraries of high Kolmogorov complexity and model a Bernoulli process, and in another partitioning may result in itineraries of much lower complexity and model much more regular, history-dependent, stochastic processes.

([Win97], p. 309)

This dependence entails an inherent asymmetry between deterministic dynamics and its alternative: while the former can *in principle* be cooked up to mimic the latter, such an effort always requires more physical resources (think of the ancilla field above, formally represented as a Hilbert space with a higher dimension, or, in the classical case, of a finer grained resolution of phase space that removes the alleged indeterminism). But now one must introduce another level of complication to the empirical indistinguishability claim, that rests on the physical possibility of concocting deterministic dynamics to model stochastic behavior, and not only on the mere logical possibility of doing so. In other words, if we demand that the additional mathematical structure, required to turn indeterministic dynamics into deterministic dynamics, has a physical counterpart, then the question of the physical resources, required for the realization of this physical counterpart, suddenly becomes relevant.

And so in order to break the alleged underdetermination between determinism and indeterminism as possible descriptions of the world, one need only assume, as the finitist does, that physical resources in the world, required, for example, for a higher resolution of the physical state and its dynamics, are bounded from above:

Some deterministic systems, when partitioned, generate stochastic processes. No one of these stochastic processes can, however, generate the deterministic flow. The deterministic flow is, if you like, a recipe for generating stochastic processes, none of which can, in return, generate its parent flow . . . Suppose that our viewer is the finest or most accurate viewer currently available. Then, do not predictions about what *would* happen *if* we had a finer viewer become empirically empty?

([Win97], p. 317)

The upshot is that whether or not there exists a bound on physical resources, and, if so, what this bound is, are both pertinent to the question of determinism. While deterministic chaos is compatible with randomness in models wherein the state space is a continuum, the same need not be so when the state space is countably infinite or finite. So long as a question of a realistic interpretation of the continuum assumed by a chaotic dynamical model remains open, the question of the ontological status of randomness in the physical system modeled will remain open too.

One could argue, of course, that for this argument about the inherent asymmetry in the mathematical descriptions of indeterministic phenomena to work, not only must the finitist have knowledge of the actual bounds, but she also must presuppose that the current state of affairs – that is, the current resolution employed in the description of the phenomenon at hand – saturates these bounds. In other words, the finitist must presuppose that the question of spatial discreteness is, in itself, an empirical question and not a metaphysical one. I completely agree, but this only means what we have hinted above, namely, that the question of spatial discreteness is epistemically prior to the question of determinism; whether or not the latter is metaphysical depends on whether or not the former is so.

3.5 Enough with the "isms"

In this chapter we investigated the philosophical price of finitism, both as a meta-physical thesis that rejects actual infinity, and as an epistemological thesis that is motivated by the precedence of measurable quantities.

We concluded that, on both counts, this price is not as intolerable as some would like us to believe. Finitism might blur the distinction between epistemology (what we know) and ontology (what is), but this need not entail that it cannot distinguish between quantum and classical probabilities. Finitism may be motivated by operationalism, but this need not turn the question of the applicability of discrete geometry to the physical world into a purely metaphysical one.

The merit of these philosophical arguments becomes apparent when one trans-lates them into methodological language. It now becomes much more evident, for example, that, either as a metaphysical thesis, or as an epistemological methodol-ogy, the application of finitist ideas in physics must be flexible enough to allow spatial discreteness to yield *in principle* novel predictions, while accounting for all the phenomena that modern physics has been explaining so successfully with the continuum. The rest of the book will thus focus on the attempts to introduce spatial discreteness into physics, either as an ontological feature of spacetime, or as an epistemological constraint on measurement resolution, and on the hope of making the question of fundamental length genuinely empirical.

4

Electrodynamics, QED, and early QFT

4.1 Outline

The first thing one notes when one turns to physics is that, remarkably, while the microscopic regime is equipped with several scale factors, or characteristic lengths, there is no fundamental constant of length dimension.

Classical mechanics, using the dimension of length l, mass m, and time t, has one universal constant G, namely Newton's gravitational constant. The theory is scale independent (the laws scale over an arbitrary range) as long as l, m, and t change so as to satisfy the constancy of G. In a similar way, classical electrodynamics has one universal constant, the velocity of light c, and the theory is scale independent in the above sense if l and t change in a way that preserves c. Only with the classical theory of the electron did physics have sufficient universal constants, those of charge e, mass m, and velocity c, to form a quantity with the dimension of a length, namely the electron radius $r_0 = e^2/mc^2$, which fixes the scale of the theory [McK60]. Adding a fourth constant, Planck's $\hbar$, to e, m, and c, quantum electrodynamics (QED henceforth) allowed a number of new ways to form characteristic lengths, for example, the Bohr radius $a_0 = \hbar^2/me^2$, or the Compton wavelength $\lambda = \hbar/mc$, which, respectively, fix the scale for (i) atomic dimensions and (ii) the relativistic behavior of the electron.

But while for every mass m_i there is a corresponding characteristic length $l_i = \hbar/m_i c$, whose physical significance is that the relevant mass cannot be confined to a region below this length (otherwise the number of particles becomes indefinite, as the energy necessary for the particle's confinement would precipitate into new particles), no universal length constant appears in microphysics.

Are there any theoretical motivations for introducing such a magnitude, and if so, how should this universal constant be incorporated into our physical theories?

From a strictly historical perspective, while the notion of fundamental length had been entertained earlier by a minority of physicists as an alternative to the continuum in the classical theory of the electron [KC94, Roh07], it became a major issue of discussion in mainstream theoretical physics only around the 1930s. It was in those years that several physicists answered the first question in the positive, and further suggested either to modify the kinematics of our theories, or to modify the dynamics appropriate for classical or quantum field physics at small distances. Depending on the physicist's taste, these modifications would then be interpreted either ontologically as representing discretization of the geometrical structure of space(time), or epistemically as representing limitations on spatial measurement resolution which indicated a breakdown in the applicability of the theory.

These were years of crisis for quantum theory, as the attempts to unify it with the special theory of relativity had reached what seemed to be an impasse [Gal83, Rue92]: the logical consistency of QED was under attack [LP31, Jac10], and the discoveries of the neutron (1932) and the meson (1937, today's pion) called into question the empirical motivation for the new theory. In addition, physicists encountered a severe problem when several quantities appearing in the formalism diverged.

Some of these infinities, for example, the self-energy of the electron (the energy of the electron in its own electromagnetic field), arise already in a classical context and result from the conceptual tension between the notion of a dimensionless point-particle and the notion of a continuous field. Others, such as those arising from the attempt to account for wavelengths of any shortness, i.e., to quantize a field with infinite degrees of freedom, or from the behavior of vacuum polarization, are germane to quantum theory and have no counterpart in classical physics, although, as we shall see below, they too are a manifestation of the problematic concept of "point-coupling."

The crisis was partially resolved in 1933 when Rosenfeld and Bohr [BR33] succeeded in addressing the logical concerns, demonstrating that the consequences of the new QED were consistent with our best possible measurements of field quantities. But the problem of divergent quantities persisted, and trying to overcome this, physicists sought a remedy in the notion of a fundamental length.

In physics an integral diverges because of contributions of objects with very high or very low energies (or equivalently, when describing phenomena at very short or very long distances). The former is often called *ultraviolet*, after the famous "ultraviolet catastrophe" of blackbody radiation in the nineteenth century, that ultimately led to the development of quantum mechanics [Kuh78]. The latter, which we will ignore in our discussion, is called *infrared*.

The elimination of divergent magnitudes, or singularities, is commonly done in mathematical physics by imposing a cutoff, or a bound, on the integral, thereby replacing the infinite magnitudes with finite ones. This trick is used, for example, to tame the singularities that threaten Newtonian gravity in collisions of point-particles whose center-of-mass-distance vanishes [Ear86], and is apparent in Planck's treatment of blackbody radiation. It was thus only natural to adopt it in the attempts to solve the problem of divergences that saturated field theory – whether classical or quantum. The cutoff on high momentum transfer, in this context, was a candidate solution, and, as we shall see, it is here where the notion of fundamental length entered the discussions in modern physics.

In this chapter we shall survey some of these attempts. Instead of an exhaustive account, only a few selected milestones will be presented, that best emphasize the conceptual difficulties that surround the notion of fundamental length. As it turns out, these difficulties, and the methods that were used to tackle them, have resurfaced in other domains of theoretical physics, particularly in the current struggles to construct a quantum theory of gravity.

4.2 Classical electrodynamics

The clash between the continuous and the discrete, described from the mathematical and philosophical perspectives in the preceding chapters, became evident in theoretical physics with the attempts at reconciling field theories with the notion of a particle.

Prima facie there are two ways to conceive of a particle: either as a singularity (a dimensionless point) or as a finite extended object. The first alternative endorses actual infinity; the second rejects it. Both conceptions, as we shall see, have generated insurmountable problems in the classical case, where physicists have tried to generalize Maxwell's equations of classical electrodynamics in order to consider the presence of charged particles.

4.2.1 The extended electron

The description of macroscopic electromagnetic phenomena was formalized mathematically by J. C. Maxwell around 1865. Almost thirty years later, Lorentz constructed a microscopic theory of electromagnetic phenomena by using Maxwell's equations and adding to them an expression for the force which a charged particle experiences in the presence of electric and magnetic fields.[1] The microscopic theory described matter in terms of its (charged) atomic building blocks, ions and electrons. In this sense it was no different than its predecessor theories of particle electrodynamics, revived by, for example, Helmholtz and Thomson. The theory drew its success, however, from the proof, first provided by Lorentz, that

---

[1] Lorentz' first paper on the subject appeared in 1892. A comprehensive presentation of the theory was published as a book [Lor16] more than a decade later.

the macroscopic theory can be reduced to the microscopic theory by a suitable averaging process over the motion of these individual building blocks. In addition to this success, and the ability of the theory to account for phenomena such as the Zeeman effect, the dispersion of light, and the propagation of light in moving bodies, the fundamental interaction in the theory differed significantly from its predecessors; it was not a direct interaction between the charges, but a mediated one, via the electromagnetic field.

The classical theory of charged particles thus emerged as a hybrid theory of particles and fields. But Lorentz was more ambitious. He wanted to describe the inner structure of an individual electron, and to establish it as a purely electromagnetic object. In particular, the mass of the electron was to be the mass equivalent of its electromagnetic energy. This project, carried out mostly by Max Abraham, was not very successful.

Abraham first chose to regard the electron as a rigid charged sphere, with a finite radius R, and with a spherically symmetric charge distribution. The starting point of the theory is the Lorentz force and the microscopic equations which, when averaged, produce Maxwell's equations. When the Lorentz force is used to describe the action which the electron's own electromagnetic field exerts on its source, the electron, the following equation of motion is obtained in ascending powers of R (the so called "structure terms"):

$$\mathbf{F} = \frac{4}{3}m\mathbf{a} - \frac{2}{3}(e^2/c^3)\dot{\mathbf{a}} + \underbrace{(\ldots)R + (\ldots)R^2 + \ldots,} \tag{4.1}$$

where m and e are the mass and charge of the electron, c is the velocity of light, the acceleration $\mathbf{a}$ and the force $\mathbf{F}$ are three-vectors, and the dot indicates a time derivative. Note the factor $\frac{4}{3}$ by which the inertial term in equation (4.1) differs from the Newtonian form. A purely relativistic definition of the four-momentum would later remove this term.[2]

The Lorentz self-force $\mathbf{F}$ (the force the electron exerts on itself) is due to the external electric field $\mathbf{E}$ and magnetic flux density $\mathbf{B}$, and is a sum over the charge density ρ:

$$\mathbf{F} = \int \rho(\mathbf{E} + \mathbf{v} \times \mathbf{B})d^3x, \tag{4.2}$$

where $\mathbf{v}$ is the electron velocity. The electron's structure is therefore characterized by a charge density distribution. This quantity has to be assumed, since there is nothing in the theory which would determine it. Only the total charge e is known from experiment.

---

[2] The first suitable merger of the Lorentz–Abraham theory with relativity was done by Fermi who noticed in 1922 that Abraham's definition of momentum was erroneous. Fermi's result went unnoticed and was rediscovered several times, recently by F. Rohrlich [Roh60]. For further details see [Roh07], pp. 16–18, 123–134.

The mass m is completely electromagnetic and is therefore given by the energy W of the electric field when the particle is at rest,[3]

$$m_e = \frac{W}{c^2} \cong \frac{e^2}{R} \frac{1}{c^2}.$$

(4.3)

The Lorentz–Abraham theory, however, had only a limited applicability. First, equation (4.1) is not a Newtonian equation of motion as it contains derivatives of the acceleration to all orders. This means that even if, as later was assumed, $R \to 0$, it remains a third order differential equation and so initial position and velocity are insufficient to determine the motion. This problem leads to a whole family of physically meaningless "runaway" solutions in which the electron has a velocity which increases asymptotically to the velocity of light irrespective of the applied forces, and in particular even when no forces are acting.

Second, the theory thus far is non-relativistic, and cannot be made consistent with the Lorentz transformations in a straightforward manner. Since "length" is not a relativistically invariant concept, a sphere in one Lorentz frame will not be a sphere in another Lorentz frame.[4] When the meaning of the Lorentz transformations became better understood (due primarily to Einstein and his kinematical explanation of the Lorentz–FitzGerald contraction), Abraham's rigid electron had to be abandoned in favor of a "compressible" oblate spheroid [JM07].

A third difficulty lies in the "structure terms." These terms depend on the charge distribution and the radius of the electron, which means that the whole dynamics becomes explicitly dependent on the electron structure. One can eliminate this dependence only by eliminating the electron structure altogether. If we assume the particle to be a point-particle it will obviously have no structure: when the electron radius $R \to 0$ the "structure terms" all vanish. This move, however, produces a new difficulty. According to equation (4.3), if $R \to 0$, then $m_e \to \infty$, i.e., the electron mass becomes infinitely large. This is the famous problem of the electron self-energy, which carries over to quantum theory. It is fair to say that a completely satisfactory solution of this problem is not known even today. A "best available" solution is provided by the renormalization procedure that will be discussed in Section 4.4. In any case, this procedure no longer allows for a completely electromagnetic mass, as Lorentz and Abraham had hoped.

---

[3] This equation assumes a uniform spherical charge distribution. Note that a different charge distribution will not modify the equation as the energy will be modified by a factor of order 1. For this reason, the quantity $r_0 = e^2/m_0 c^2$ has been designated as the classical electron radius when m_0 is the electron rest mass; it gives the correct order of magnitude of the radius for *any* charge distribution, see [Roh07], p. 125.

[4] This tension with Lorentz transformations, first noticed by Abraham [Abr04], was initially brushed away as one could still uphold a rigid model as long as there was no motion with respect to the ether.

The fourth and final difficulty is not apparent from the above equations. According to Coulomb's law, the various parts of a charged sphere must repel one another, which means that the extended electron is unstable: any finite charge distribution of only one sign would explode. Consequently, a completely electromagnetic theory of the electron could not even account for its existence. A solution was proposed by Poincaré [Poi06] who showed that an attractive and consequently non-electromagnetic force (what came to be known as "Poincaré pressure") can always be added to the theory so as to just balance the stresses and establish stability. While this solution was considered "ad hoc," unsuitable in a fundamental theory, it nevertheless made the theory consistent with the Lorentz transformations.

As a consequence of these difficulties, around 1910 the Lorentz–Abraham extended electron theory had arrived at an impasse: while it appeared to agree with the newly born special theory of relativity, the idea of a purely electromagnetic electron had to be abandoned, and cohesive forces had to be postulated to ensure stability of any finite size electron. On the other hand, dimensionless point-electrons could not yield a meaningful theory because their self-energy diverged and, consequently, the "structure terms" had to be taken seriously.

4.2.2 Dealing with infinities

Lorentz and Abraham hoped to obtain the electron equation of motion from (1) Maxwell–Lorentz field equations, (2) the Lorentz force law, and (3) the assumptions on the structure of the electron. In the three decades that followed, numerous attempts were made to modify each of these three ingredients while ensuring Lorentz invariance of the theory as a whole. Two such theories were proposed first by Born and Infeld and later by Dirac.

Born and Infeld

One of the first attempts to obtain a consistent classical equation of motion for the electron was due to Mie [Mie12], who suggested modifying the field equations to ensure the stability of the electron without the introduction of additional cohesive non-electromagnetic forces. These modifications led to non-linear equations, and to further mathematical complications. Consequently, the program was abandoned for almost two decades, when it was picked up by Born and Infeld [Bor33, BI34], in a completely different context.

Apart from a desire, similar to that of Lorentz and Abraham, for a monistic, electrodynamic, description of nature, Born and Infeld cited as their motivation for a modified field theory the problem of self-energy of the electron, which, as mentioned above, persisted in the newly developed relativistic version of quantum theory. They preferred, however, to tackle the classical field theory first, before quantizing it, for the following reason:

> In all these cases there is sufficient evidence that the present theory (formulated by Dirac's wave equation) holds as long as the wave–lengths (of the Maxwell or of the de Broglie waves) are long compared with the "radius of the electron" e^2/mc^2, but breaks down for a field containing shorter waves. The non-appearance of Planck's constant in this expression for the radius indicates that in the first place the electromagnetic laws are to be modified; the quantum laws may then be adapted to the new field equations. . . . The purpose of this paper is to give a deeper foundation of the new field equations on classical lines, without touching the question of the quantum theory.
>
> *([BI34], p. 426)*

It may be instructive to go into some details here, as the strategy Born and Infeld adopted found its way into current attempts at introducing a fundamental length into theories of quantum gravity.

Their suggestion was to replace the Lagrangian of Maxwell's theory

$$L = \frac{1}{2}(\mathbf{H}^2 - \mathbf{E}^2), \tag{4.4}$$

where $\mathbf{H}$ and $\mathbf{E}$ are vectors designating the electric and the magnetic field strengths, with the new Lagrangian

$$L = b^2 \left(\sqrt{1 + \frac{1}{b^2}(\mathbf{H}^2 - \mathbf{E}^2)} - 1 \right), \tag{4.5}$$

which reduces to the former for a small b^{-1}. This modification amounts, physically, to the introduction of a cutoff, or an upper bound, on the allowed energy of the field. Born and Infeld compared this move to the similar move made by Einstein, when he introduced an upper bound on velocity, thereby replacing the Newtonian action of a free particle $\frac{1}{2}mv^2$ with the relativistic expression $mc^2(1 - \sqrt{1 - v^2/c^2})$.

From a mathematical perspective, the introduction of the upper bound deforms the spatiotemporal transformation group that leaves the Lagrangian invariant. As in the case of the shift from Galilean to Lorentzian transformations, the new transformation group that leaves the new Lagrangian (4.5) invariant is larger than the original (Lorentzian) group that leaves the Maxwell Lagrangian (4.4) invariant.[5]

The constant b is of the dimension of field strength (Born and Infeld called it "the absolute field"). Since the theory allows one to obtain a finite value for the

---

[5] In identifying the mathematical conditions that can constrain a general transformation group of this sort, namely that the field is represented by an anti-symmetric tensor, and that the Lagrangian is the square root of its determinant, Born and Infeld followed Eddington, who suggested this constraint a decade earlier, in *The Mathematical Theory of Relativity*, see [Edd23] §48, §101.

mass or energy of the electron with a definite numerical factor, both the strength of the absolute field and the size of the electron radius r_0 can be computed ([BI34], p. 446):

$$r_0 = 2.28 \times 10^{-13} \text{ cm} \tag{4.6}$$

and

$$b = \frac{e}{r_0^2} = 9.18 \times 10^{15} \text{ e.s.u.} \tag{4.7}$$

Due to the enormous strength of the field (or, equivalently, the small radius), Maxwell's equations apply in all cases, except those where the inner structure of the electron is concerned (i.e., when the field strength is of the order b, or when the distance or wavelength is of the order r_0), which also means that the theory is indifferent to the actual question of whether the electron is a point or an extended particle:

Our theory combines the two possible aspects of the field; true point charges and free spatial densities are entirely equivalent. The question whether the one or the other picture of the electron is right has no meaning.

([BI34] p. 445)

The Born–Infeld theory modifies Maxwell's equations only for field strengths of order b or larger. Moreover, the theory is relativistic (although its transformation group is larger than the Lorentz group), and it completely removes the problems that saturated the Maxwell–Lorentz theory:

The Born–Infeld electrodynamics is completely satisfactory as a classical theory. It avoids all the troublesome infinities associated with point charges in the Maxwell theory. Also, it gives an expression for the energy density . . . that is positive definite . . . It follows that there cannot be any runaway solutions of the equations of motion for a charged particle (in which the particle continually accelerates in the absence of an external field), such as occur with the Maxwell–Lorentz theory.

([Dir60], pp. 42–43)

Nevertheless, the theory has remained outside mainstream physics (until recently when it was "rediscovered" as a possible limit to string theory [CM98]), for the following reasons.

First, the non-linear nature of the theory is evident, and is a source of great mathematical complications. Second, essential modifications to Maxwell's theory only "kick in" in domains that were beyond resolution capabilities at that time, and so, for many, the motivation for introducing an upper bound on energy (or a notion of a fundamental length) was unjustified:

> ... modifications of these sort are legion, and such attempts therefore remain unsatisfactory as long as there is no experimental evidence in favor of any of them.
>
> *([Roh07], p. 19)*

Finally, the most serious drawback of the theory was the fact that it was not quantizable: it could not be quantized with contemporary rules of quantization, even in the absence of charges. This drawback was evident to Born [Bor49] and was also emphasized by Dirac ([Dir60], pp. 42–43).

Despite its great theoretical benefits, the theory was thus rejected for lack of phenomenological justifications, as well as for the inability to quantize it.

Dirac

With the development of quantum mechanics more attention was given to the problem of self-energy, and attempts were made to confront it. But many physicists were unconvinced and supported the opposite strategy, voiced by Born and Infeld, namely that only after obtaining a satisfactory classical theory of the electron, independent of superfluous considerations about its structure, should quantization be imposed:

> One may think that this difficulty [i.e., the divergence of self-energy – AH] will be solved only by a better understanding of the structure of the electron according to quantum laws. However, it seems more reasonable to suppose that the electron is too simple a thing for the question of the laws governing its structure to arise, and thus quantum mechanics should not be needed for the solution of the difficulty ... Our easiest path of approach to it is to keep within the confines of the classical theory.
>
> *([Dir38], p. 148)*

In 1933, Wentzel discovered that a convergent result for the self-energy in a Maxwell–Lorentz theory with a point-electron can be obtained by a careful limiting process [Wen33]. Five years later, Dirac [Dir38] presented a relativistic generalization of the Lorentz equation (4.1) when applied to a point-electron ($R \to 0$). To avoid the divergences, Dirac proposed a process by which the electromagnetic mass, while infinite when $R \to 0$, is lumped into the total mass of the electron to yield its observed mass, thus offsetting its divergent nature:

> If we want a model of the electron, we must suppose that there is an infinite negative mass at its centre, such that, when subtracted from the infinite positive mass of the surrounding Coulomb field, the difference is well defined and is just equal to m. Such a model is hardly a plausible one according to current physical ideas, but, as discussed in the introduction, this is not an objection to the theory provided we have a reasonable mathematical scheme.
>
> *([Dir38], p. 155)*

Apart from this idea of negative energy, the theory still suffered from the problem of runaway solutions, and, furthermore, from a failure of causality. The reason for the latter was that the new Lorentz–Dirac equations of motion still contained a third time derivative of position, and allowed for an electron disturbed by a force to accelerate (and radiate) before the applied force had reached it. Dirac's own solution to this problem is telling:

> The behaviour of our electron can be interpreted in a natural way, however, if we suppose the electron to have a finite size. There is then no need for the pulse to reach the centre of the electron before it starts to accelerate. It starts to accelerate and radiate as the pulse meets its outside. Mathematically, the electron has no sharp boundary . . . but for practical purpose it may be considered to have a radius of order a^{-1}, this being the distance within which the pulse must arrive before the acceleration and radiation are applicable.
>
> *([Dir38], pp. 159–160)*

The introduction of the electron radius would thus solve the causality problem by "pushing it," as it were, into the electron inner structure.[6] The price, however, is that failure of causality (manifest in, e.g., instantaneous signaling) is possible *within* the interior of the electron. According to Dirac, this scale is thus a region of failure not of the electromagnetic field equations, but of some of the elementary properties of space and time. Note that despite this departure from relativity, the equations of motion in the theory were still Lorentz invariant above that scale.

With the development of QED, interest shifted to the new theory and the problems it was facing. New particles were discovered, and the classical theories of the electron with their limited domain of applicability were forgotten, except by general relativists who were looking for unified theories that combined gravitational and electromagnetic interactions. But the attempts of Lorentz, Abraham, and Dirac to solve the conceptual and the mathematical problems of the classical theory extended electron found their way into the new domain of QED, to which we now turn.

4.3 From QM to QED

The history of the years that led to the birth of QED and ultimately to quantum field theory has been investigated thoroughly by many experts (see, e.g., [Wei77, Cas81, Gal83, Dar86a, Dar86b, Gal87, BR91, HK91, Kra92a, Rue92, Mil94, Sch94, Car96] and references therein).[7] The story is well known: the divergences that plagued the theory were eliminated in a suitable way only after the discovery

---

[6] In this sense, it is misleading to call Dirac's theory a point-electron theory (as, e.g., in [CS93], p. 37 or [Roh07], pp. 20–21).

[7] The amount of literature on this subject is overwhelming. [HK91] alone contains almost 500 items of bibliography!

of renormalization, which relieved theoretical physics – if only temporarily – from the attempts (and the consequences thereof) at introducing fundamental length into the formalism. As Weinberg put it, "things changed only just enough so that they could stay the same" ([Wei77], p. 18).

In the following section we shall discuss two types of such attempts. The first type involves treating fundamental length as a kinematic concept, i.e., as a limitation imposed on the measurement resolution of field operators. The second type involves introducing fundamental length as a dynamical concept, i.e., as a limit on possible interactions via high momentum cutoffs. These two approaches are, of course, interconnected, and both signify a limitation on the applicability of QED: on the one hand, measurements require interactions, and so the inability to resolve certain wavelengths follows from energy bounds one imposes on possible interactions; on the other hand, limitations on measurement resolution imply that interactions above a certain threshold cannot be described by the theory.

4.3.1 Fundamental length as a kinematic concept

Quantum mechanics (QM) introduced discreteness into the description of nature, at least as a limit on measurement resolution, with Heisenberg's uncertainty principle [Hei27]:

$$\Delta p \Delta q > \frac{1}{2}, \tag{4.8}$$

where Δp and Δq are the spreads of the probability distributions in position and momentum, respectively, and for simplicity we set $\hbar$ to 1 and confine the discussion to one dimension. Heisenberg's initial explication for the uncertainty relation was the discretization of phase space – an idea that goes back to Boltzmann and his ergodic hypothesis [Gal99] – into cells of size h. In a letter to Pauli from November 15, 1926, Heisenberg already speculates about a relation such as (4.8), and also mentions a discussion he had with Bohr on the idea that the essential discreteness of quantum theory is a clue to the possible discreteness of the spacetime metric,[8] which implies the impossibility of joint measurements of position and momentum with arbitrary degree of accuracy ([Mil94], p. 14).

Attempts at unifying QM with the special theory of relativity by, for example, quantizing the electromagnetic field, generated several problems. Some of these, for example, the divergence of the self-energy of the electron, were imported from the classical domain. Others, however, were germane to the uncertainty principle.

---

[8] The same idea led Heisenberg a few years later to the notion of "lattice world" [Hei30b, CK95]. We shall say more on this intuition, that depicts the uncertainty relations as emerging from spatial discreteness, in Section 5.6 below.

The debate that ensued on the consistency of the uncertainty relations when applied to field measurements is crucial to our story.

Heisenberg and Pauli

In 1929 Heisenberg and Pauli published two long papers in which they developed a theory of the interaction of light and matter by quantizing Maxwell's equation of the electromagnetic field [HP29]. One of the difficulties that emerged from their papers was the problem of the self-energy of the electron. At first they discarded it as an irrelevant additive infinite constant, just like the zero-point energy of the vacuum (see below), but in the second paper they admitted that this move restricts the domain of applicability of the theory. Despite this shortcoming, Heisenberg and Pauli were confident that any future correct theory of field quantization, even quantization of the gravitational field, "... should be feasible without any new difficulties by means of a formalism completely analogous to the one used here" ([Mil94], p. 34, fn. 14).

Soon after, the presence of infinities was confirmed in calculations of the electromagnetic self-energy of a bound electron by Oppenheimer, and of a free electron by Waller. Many physicists tried to grapple with this problem (see [Rue92] and references therein). By the mid-1930s it became clear that it was the local coupling of the charge-current density to the electromagnetic field – corresponding in the classical theory to the coupling of a point charge to the field – that was responsible for giving arbitrary short wavelengths a divergent role in calculations of both the self-energy and the vacuum polarization ([Sch94], p. 87). These divergences led to the contemporary belief that, similarly to its classical twin, the newly born QED could not account for the inner structure of the electron. This failure, as we have seen in the case of Dirac's classical theory of the electron, could also be interpreted as a drastic change in some of the essential properties of spacetime at short distances.

Another difficulty emerged from the generalization of the uncertainty principle to field quantities. Heisenberg and Pauli observed that the quantum conditions give commutation relations between positions and momenta at two different points at the same time and are therefore non-relativistic in form. To restore relativistic invariance, field operators at points connected by finite spatial displacement (space-like separated) must always commute, so one can measure one of them without disturbing the other, to the extent that one measurement does not change the statistics of the other. This condition was later called "microcausality."

But while quantization of the electromagnetic field could be reconciled with relativistic invariance with the help of the microcausality condition, it generated another problem, commonly known as the "zero-point energy."

The uncertainty principle implies a limit on the accuracy with which one can measure certain field quantities (just as it restricts the accuracy of measurements of position and momentum coordinates of a particle). The "price" for infinitely precise measurements at a given point in space is that the uncertainties in momentum (and hence in energy) in the field variables become *infinite* at that point. Its apparent quantum origin notwithstanding, this divergence could be dealt with in a similar way to the classical self-energy, namely, by introducing a finite extension to the formalism. This recognition, that in quantum field theory only *smeared* operators – operators suitably averaged over small but extended spatiotemporal regions – make sense, was a result of a famous debate on the measurability of field quantities [Jac10].

Landau and Peierls

While the uncertainty relations refer to measurements of conjugate observables, in 1931 Landau and Peierls [LP31] argued that not even a *single* field component could be measured with arbitrary accuracy:

> ... no predictable measurements can exist for the fundamental quantities of wave mechanics (except when these quantities are constant in time, and then an infinitely long time is needed for an exactly predictable measurement) ... It is therefore not surprising that the formalism leads to various infinities; it would be surprising if the formalism bore any resemblance to reality. ... In the correct relativistic quantum mechanics (which does not yet exist), there will therefore be no physical quantities and no measurements in the sense of wave mechanics.
>
> *([LP31], pp. 474–475)*

The consequences of this state of affairs, if true, were that QED was epistemically incoherent: the theory itself precluded its own verification.

To substantiate their harsh verdict, Landau and Peierls argued the following (for a comprehensive analysis see also [Kal71]): consider the measurement of an electric field component in the x direction E_x averaged over a time interval T, carried out by determining the momentum transfer from the field to a charged test body with a charge q during that time,

$$P_x^{(2)} - P_x^{(1)} = q\bar{E}_x T, \qquad (4.9)$$

where the momentum before and after T is denoted by $P_x^{(1)}$ and $P_x^{(2)}$, respectively.

Since momentum measurements are restricted by the uncertainty principle, assuming they are done with accuracy ΔP_x, we must allow an uncontrollable latitude in the position of the test charge $\Delta x \sim \hbar/\Delta P_x$, thereby restricting the accuracy of the field component measurement:

$$\Delta \bar{E}_x \sim \frac{\hbar}{q \Delta x T}. \tag{4.10}$$

Prima facie this uncertainty may be made arbitrarily small for a given Δx by choosing a sufficiently large q. But Landau and Peierls noted that an increase in the charge of the test body would increase the contribution of the field *reaction*, and so the correct calculation of the error in momentum ΔP_x during the time interval Δt taken up by the first and second momentum measurement (in which the test body – considered as a point charge whose mass is large enough to neglect additional acceleration – suffers an uncontrollable energy and momentum loss through radiation) is

$$\Delta P_x > \frac{h}{c \Delta t} \sqrt{\frac{q^2}{\hbar c}}. \tag{4.11}$$

Adding this uncertainty into the initial uncertainty of the field component measurement without field reaction (4.10) leads to a minimum uncertainty that, even after variation of q, remains finite, and vanishes only for (unphysical) infinitely long measurements:

$$\Delta \bar{E}_x^{\min} > \frac{\sqrt{\hbar c}}{(cT)^2}. \tag{4.12}$$

Thus, according to Landau and Peierls, since the emitted radiation would interfere with the field of the test body that was being used to observe the original field, it is not actually possible in practice for any kind of device to measure even a single component of that field by itself beyond a certain limit of accuracy.

Landau and Peierls' argument, as well as the response it generated which we shall discuss below, are motivated by the so called "disturbance" view of the uncertainty principle. This view has been criticized by many (e.g., [BR81]), as inadequate, as it appears to lend support to an epistemic, "ignorance," interpretation of QM, in which classical values exist but we are ignorant thereof because of the inherent limitations on measurements. I shall say more on this issue in Section 5.6. For now, let me just note that, on the finitist and operationalist view presented in Section 3.2.4, quantum probabilities result from an inherent limitation on measurement resolution, so the metaphysical question of properties having sharp values without measurement is sidestepped as moot, since the difference between quantum and classical probabilities is seen as quantitative and not as qualitative. The problem with Landau and Peierls' argument is not their reliance on the "disturbance" view; rather, from the perspective we have been pursuing here, the problem with Landau and Peierls' argument is that it relies on the idea that field quantities should be measured at dimensionless points.

Bohr and Rosenfeld

Bohr disagreed with Landau and Peierls' result, and in 1931 he and Rosenfeld [BR33] set out to restore epistemic coherence to the newly born theory of QED. Two assumptions underlie their response to Landau and Peierls. First, field components at dimensionless points are a mere idealization without immediate physical meaning; the only physically meaningful statements of the theory concern averages of field components over extended spacetime regions [Ros55]. Second, due to the smallness of the fine-structure constant $e^2/\hbar c \approx 1/137$, in the analysis of the measurement process it is admissible to neglect all consequences of the inner structure of the test body:

So long as we treat all sources of electromagnetic fields as classical distributions of charge and current, and only quantize the field quantities themselves, no universal scale of space–time dimension is fixed by the formalism. It is then consistent to disregard the atomistic structure of test-bodies . . .

([Ros55], p. 72)

These two assumptions allowed Bohr and Rosenfeld to conceive of measurement procedures that completely compensate for the field reaction (as long as one neglects the extended body inner structure), and thus lead to an uncertainty in the field component measurement of the form:

$$\Delta \bar{E}_x \sim \frac{\hbar}{\rho \Delta x V T} \tag{4.13}$$

where V is the spatiotemporal volume filled by the extended test body, and ρ is its charge density (here again the test body is assumed to be sufficiently heavy as to reduce its displacement during the measurement interval). This expression allows one to reduce arbitrarily the uncertainty in the measurement of a single field component by increasing the charge density without worrying about field reaction.

After demonstrating, with the help of ingenious thought experiments involving constructions such as mirrors and springs, that it is possible in principle to construct devices that would measure one component of the field, averaged over a finite volume (or over a finite time), to any degree of accuracy,[9] Bohr and Rosenfeld further proved the consistency of the uncertainty relations implied by the commutation rules of the electromagnetic field components.

The same methodology guided Bohr and Rosenfeld several years later in their treatment of measurability of charge and current densities [BR50]. This piecemeal methodology of quantization – first quantize the radiation field, and only then the fields associated with the charged particles – was, of course, in accord with Bohr's

---

[9] Although fully consistent, the measurement procedure described by Bohr and Rosenfeld is rather curious and unlike anything an experimentalist would design in practice [Pei63].

insistence on the primitivity of classical concepts in interpreting the formalism of quantum theory. In this later treatment they noted ([BR50], p. 798, [Ros55], pp. 83–84) that while their analysis allows one to resolve shorter and shorter wavelengths without losing measurability of charge density, the Compton wavelength of the electron marks the limit of applicability of their idealizations – those idealizations that allow them to conceive of measurement procedures that compensate for field reactions. Below this characteristic length, short range forces dominate that have no analog in classical electrodynamics, hence presumably a different theory than QED is required.

4.3.2 Fundamental length as dynamical concept

In the two decades or so between 1929 and the rise of renormalization, several abortive attempts were made, motivated in great part by the divergence problems, in which the notion of fundamental length also played a crucial dynamical role. Some of these attempts have already been discussed by historians (e.g., [CK95, Kra95]), and some are less known.

Heisenberg's lattice world and the quest for universal length

In an introduction to a paper published in 1930, Heisenberg [Hei30b] discussed a possible solution to the problem of the self-energy of the electron, that requires a radical modification of current ideas:

> If one decides favorably of such a basic modification of quantum theory, it first seems necessary to introduce the radius r_0 more or less in such a fashion that one divides space into cells of finite size r_0^3, and that one replaces the present differential equations by difference equations. In any case, the self-energy of the electron would be finite in such a lattice world.
>
> *([Hei30b], p. 121)*

Preceding this paper was a correspondence with Bohr (see [CK95], pp. 597–598) where Heisenberg repeats his intuition about the discreteness of space that underlies the uncertainty relations,[10] which may help to solve the divergences he and Pauli [HP29] had encountered in their theory.

Heisenberg's insight ([Kra95], p. 410) was to replace the differential quotients in the dynamical equations of motion with difference quotients, according to

$$\frac{\partial \psi}{\partial x} \rightarrow \frac{(\psi_{n+1} - \psi_n)}{a} \tag{4.14}$$

---

[10] A similar idea was voiced in 1928 by Ruark [Rua28] and Flint [Fli28].

and

$$\frac{\partial^2 \psi}{\partial x^2} \rightarrow \frac{(\psi_{n+1} - 2\psi_n + \psi_{n-1})}{a^2}, \tag{4.15}$$

where a is the fundamental length (which Heisenberg took to be $\hbar/Mc$, where M is the proton mass).

When plugged into the Klein–Gordon equation, this modification yields

$$-\left(\frac{\mathbf{E}}{c}\right)^2 + \left(\frac{\hbar}{ia}\right)^2 [\psi_{n+1} - 2\psi_n + \psi_{n-1}] + m^2 c^2 \psi_n = 0. \tag{4.16}$$

In his letter to Bohr, Heisenberg shows how he could obtain an energy spectrum where the electron and the proton both appeared as solutions to the same eigenvalue equation.

Heisenberg's wild guess about the composition of the nucleus was soon proven wrong [Bro71], but the technical problems of his proposal – of which he was aware – are important to our discussion, as they reappear in many of the attempts of his contemporaries to introduce a fundamental length to QED and, furthermore, in many of the current theories of quantum gravity.

In his letter to Bohr, Heisenberg immediately recognizes the inherent problems of his "lattice world": breakdown of Lorentz invariance and of space isotropy, and, above all, failure of conservation of energy, momentum, and charge, which meant that while at the atomic scale all these (well confirmed) features hold approximately, at the nuclear scale they break down. In the published version of his paper, Heisenberg repeats his concerns, and admits renouncing the whole idea:

Although such a lattice world otherwise also has remarkable properties, one must nevertheless consider that it leads to deviations from the present theory which experimentally are not probable. In particular, the statement that the smallest length exists is no longer relativistically invariant, and no way is presently known to harmonize the requirement of relativistic invariance with the fundamental introduction of a smallest length.

([Hei30b], p. 121)

Before we go on to discuss these problems, let us just mention Bohr's response to what Heisenberg himself presented as a "radical," and "completely mad," idea.

Bohr, apparently, was well aware of the intuition regarding spatial discreteness as a limit on position measurement and, at first, was quite skeptical, as these limitations would interfere with "the beauty and consistency of the theory [of relativity] to far great an extent" ([CK95], p. 603). After reading Heisenberg's letter, however, he seems to have changed his mind, but although more sympathetic, he nevertheless objected to the whole notion of a lattice world. Two problems troubled Bohr: (1) the non-conservation of charge, which was difficult to reconcile with the correspondence principle (as conceptions of charge and radiation were, according to him, classical, hence to be found directly in classical electrodynamics), and (2) the intuition that the notion of a smallest length is inconsistent with length contraction.[11]

Heisenberg agreed with Bohr's criticism, and so, almost a month after its conception, the idea of "lattice world" came to a halt, and apart from the above mentioned introduction to [Hei30b], had no other reference. Yet the idea did not die. In that year it was adopted by the young Soviet physicist Matvei Bronstein, who would develop it into an attempt at quantizing space, and would try to circumvent the problems encountered by Heisenberg ([GF94], pp. 83–121; see Chapter 5). Moreover, Heisenberg himself would come back to it in one way or another in his later work.

As for the problems that the "lattice world" generated, they were the following.[12]

- A lattice structure of space immediately breaks continuous rotational and translational symmetries, Lorentz invariance included. To see this think of a cube. It has many rotational symmetries. It can be rotated by 90 degrees, 180 degrees, or 270 degrees about any of the three axes passing through the faces. It can be rotated by 120 degrees or 240 degrees about the corners, and by 180 degrees about an axis passing from the center through the midpoint of any of the 12 edges. Now think of a sphere. It can be rotated by any angle. In this sense the sphere respects rotational invariance: all directions are on a par. The cube, on the other hand, is an object which breaks rotational invariance: once the cube is there, some directions are more equal than others. In a similar vein, a lattice breaks translation invariance: it is easy to tell if a lattice is shifted sideways, unless one shifts it by a whole number of lattice units.
- Heisenberg only considers discreteness of space, and time is treated as a continuous parameter.[13] But a discretization of length alone is inherently non-relativistic: a moving rod contracts, and so the minimal length in one frame will not be so in another. This intuition, pointed out by Bohr, returned in almost all the discussions on fundamental length in years to come.
- While it does reduce to the standard classical case when the fundamental length vanishes, charge density in the lattice world becomes velocity dependent, and so is not conserved. Furthermore, this feature contains the seed of the idea of a *form factor* (see below) that later on would appear in the research program of non-local field theories, namely, that elementary particles cannot be localized to

---

[11] Note that in contrast to the case of charge, Bohr was quite flexible about violations of energy conservation.

[12] The following is a reconstruction of Heisenberg's approach in one dimension, developed in [CK95], pp. 598–601.

[13] Carazza and Kragh [CK95] note that there are reasons to believe Heisenberg also thought about discrete time. First, he uses the word "initially," which means he had in mind further development of the theory; second, in his paper Heisenberg mentions replacing the differential equations with difference equations, so the time derivative would be replaced by a finite difference.

a dimensionless point at a given moment in time, and so physical magnitudes are "smeared out" over a finite extension of space. In Heisenberg's case, the particle–field interaction in a certain point x is determined not only by $\psi(x)$ but also by $\psi(x + a)$ where a is the lattice width and ψ is the wave function of the particle. This feature, as we shall see, generated serious problems in attempts at reconciling non-local field theories with the notion of relativistic causal order.

- Finally, the Heisenberg model was inspired by the pioneering work of his student, Rudolph Peierls, in the theory of metals. Analogously to Peierls' model, the lattice breaks conservation of momentum. If, further, one introduces discretization of time, it also breaks conservation of energy.

In years to come, Heisenberg would return to the idea of fundamental length (and of fundamental time duration) in several different contexts: a letter he sent Bohr in 1935 shows he thought the Bohr and Rosenfeld solution to the Landau and Peierls challenge proved his intuition about spatial discreteness ([Kra95], p. 411), his attempt in 1936 [Hei36] to use the electron radius in Fermi's β-decay theory to eliminate divergences, and the idea he proposed in 1938 [Hei38b], motivated by Yukawa's meson theory and the phenomenon of cosmic ray showers, to construct a universal constant of length, developed in [Hei38a], where the notion appears as a cutoff on the four-momentum transfer during a collision

$$(\Delta p)^2 - (\Delta E/c)^2 \ll (\hbar/\lambda_0)^2 \tag{4.17}$$

where $\lambda_0 = \hbar/\mu c$ and μ denotes Yukawa's meson (today's pion). Heisenberg's interpretation of this cutoff was the following (translated in [Kra95], p. 415): "A process can be treated in an approximately quantitative way by the formulae of quantum mechanics if and only if at any instant of the process a time $\Delta\tau \gg \lambda_0/c$ can be given for any particle participating in the process so that during this time interval the particle undergoes only a small change in the sense of equation (4.17)." Interestingly, physical processes above this cutoff were interpreted by Heisenberg as beyond the limit of applicability of quantum theory, and not as physically impossible, which means that he now interpreted the universal length epistemically, and not as a feature of the structure of spacetime.

Heisenberg's theory of cosmic showers did not mature. Pauli disliked it, and it was further criticized by Bhabha [Bha39] who proposed an alternative explanation of cosmic showers, rejecting Heisenberg's claim that quantum mechanics would fail for energy transfers larger than $\hbar c/\lambda_0$.

Still Heisenberg would not give up. In his famous S-Matrix theory [Hei43], he speculated about the existence of an absolute length a 10^{-13} cm or an absolute time $\tau = a/c$ 10^{-24} seconds, and voiced his doubts that the usual description of a physical system with the help of a Hamiltonian function has any meaning for space and time intervals smaller than a and τ. One can only observe alterations in time intervals which are long compared with τ. The classical correspondence limit of this theory would be a non-linear field theory, similar to the one suggested by Born and Infeld [Bor53].

The physics community, however, was still unconvinced. First, as Pauli puts it in a letter to Heisenberg from 1947:

Why shouldn't every particle have its own length? Why should the length be "universal"? There certainly does *not* exist a geometrical structure of space with a smallest length introduced *a priori* independent of the material particles.

(Pauli, quoted in [Kra95], p. 419)

Second, relativistic invariance introduces, again, strange paradoxes into this theory, as the temporal order of events, and with it the cause–effect relation, break down for short time intervals. For instance, a particle may be absorbed before the creating collision has taken place.

Heisenberg's response to this worry is telling. In a rarely cited paper he analyzes the causal paradoxes that arise in his theory, and gives two arguments for why they should be tolerated:

It seems, however, questionable, whether a direct observation of the paradoxes just described would be possible, because, of course, the measuring devices themselves consist of atoms and therefore do not allow the measurement of time differences on the order of 10^{-24} sec. One could only begin to think about measurement in processes of very high total energy, in which the particles of the second group [the virtual quanta – AH] are already created with such high energy that their disintegration time is highly magnified through time dilation, in the frame of reference of the whole process, and thereby becomes measurable. Only in processes of such high energy does the fact become detectable that the blurring of the borderline between past or future . . . also occurs over greater spatial or temporal intervals along the light cone. But it is questionable even in these processes whether the apparent time reversal can be observed. For it is quite possible that the interaction at high primary energy will make sure that the energy is distributed among many secondary particles of relatively low energy; that therefore the creation of a particle of the second group with an energy that is large compared to its resting energy, is an extremely rare process, a process whose probability decreases exponentially with the energy of the particle. With that sort of low probability, very paradoxical processes occur in ordinary quantum mechanics, too. On the whole, the fact that measuring apparatuses are made up of atoms should practically forbid a direct observation of the paradoxes just mentioned. An indirect observation should naturally always be possible, insofar as the situation represented by [the equations of the theory – AH] must be gleaned from the experiments.

([Hei51b], pp. 60–61)

The point in this passage is that the causal anomalies may be unobservable *in practice* because of the macroscopic character of the measurement instruments: (1) observations of the anomalies require high energy, and (2) even if one succeeds in achieving such energies, statistical considerations of energy partition ensure that the probability of detection of these anomalies becomes exponentially small. Note that Heisenberg insists on leaving these anomalies detectable *in principle*, albeit indirectly through confirmations of other predictions of his theory.

Gleb Wataghin and the program of non-local field theories

Heisenberg's interest in the notion of fundamental length is as well known as Heisenberg himself, but in the 1930s many less known physicists were entertaining this notion as a possible solution to the divergence problems of QED (see, e.g., [AI30]). One such physicist was the Ukrainian-Italian (and later on Brazilian) Gleb Wataghin, whose short contribution to the development of the idea of fundamental length at the beginning of the 1930s spawned a research program that would later continue in the 1950s and the 1960s.

In a couple of papers from 1934 [Wat34a, Wat34b], Wataghin introduces – quite crudely – a cutoff to the momentum transfer in an elementary collision process:

$$G = \exp\left(-\frac{[\Delta p^2 - (\Delta E/c)^2]}{\Pi^2}\right), \tag{4.18}$$

where the dividend inside the exponent is the difference in momentum before and after collision, and Π^2 is a constant, for instance $c\Pi \sim 10^8$ eV.

The idea was presented as a solution to the same contemporary problem that bothered Heisenberg, namely, the self-energy of the electron. Furthermore, its direct motivation was also explicitly attributed to Heisenberg ([Wat34a], p. 92). The immediate experimental anomaly Wataghin set out to resolve was the problem of *Bremsstrahlung* (the stopping of particles in an atomic field with emission and radiation). In 1933 cosmic rays were reported well below the water level of Lake Constance. If these rays, as was generally assumed before the discovery of the muon, consisted of electrons (and not of protons, which were not observed), their great penetrating power raised a serious difficulty in the newly born QED based on Dirac's equation ([Cas81], pp. 11–17).[14] The anomaly was considered as yet another sign that QED breaks down at scales of the order of the electron radius, and is inappropriate to describe nuclear physics ([Gal87], pp. 102–110).

With the clarity of hindsight, Wataghin's idea seems completely unmotivated today. Nevertheless, it is still interesting to discuss it, as it signifies the first explicit introduction of the key ingredient in what later turned out to be non-local field theories.

---

[14] Using this equation, Heitler and Sauter [HS33] have shown that a beam of very fast electrons (with an energy $E > 200\ mc^2$) should penetrate not more than 1 m of water when all kinds of absorption processes are taken into account, see also [Bor34].

When plugged into the interaction Hamiltonian, Wataghin's cutoff makes the probability of transitions that do not concur with experiment (but that nevertheless are predicted by Dirac's equation) negligible. By estimating the limiting frequency at $137 \, mc^2/\hbar$, Wataghin implicitly introduced a characteristic length λ_0 of magnitude $\lambda_0 = e^2/m_0 c^2$, i.e., a finite electron radius. This can be seen in two different ways. First, since if $G = 1$ one gets the usual singularity, then the dividend inside the exponent in (4.18) must be finite, but if this is the case, then since momentum and position are related by Fourier transformation, the particle cannot be considered a point-particle. Second, from the Hamiltonian point of view, the interaction between particles vanishes for wavelengths shorter than the electron radius, hence one cannot resolve distances smaller than that scale.

One feature of this approach was its consistency with relativistic invariance: the cutoff was introduced into the interaction Hamiltonian in such a way that the new field operators formed a Lorentz invariant four-vector in momentum space. This formulation, which gives precedence to momentum space over position space, allows Wataghin to evade the thorny conflict of fundamental length with the notion of a dimensionless point:

Coming back to Italy [from the visit to Copenhagen, where Bohr, Pauli, and Heisenberg objected to Wataghin's cutoff – AH] after another publication [[Wat34b] – AH], I said that it is possible to formalize the Lorentz group for the energy in momentum space, and once we have momentum and energy, because of the uncertainty relations, we know nothing about space and time [position space – AH]. In this way the problem of the [dimensionless spatial – AH] point cannot even arise.[15]

([Wat75], p. 7)

It is noteworthy that even today it remains an open question whether there exists a consistent translation from momentum space to position space that can preserve both the notion of minimal length and the notion of locality (or point interaction). We shall say more on this in Chapter 8.

Another feature of this proposal was its correspondence, at the classical limit, with modified Maxwell's equations of the type suggested by Born and Infeld [Wat34b]. And there were other conceptual similarities with other contemporary approaches, for example, the electron's having no meaningful inner structure (as it cannot be resolved by any experiment). But despite these features, the suggestion was not generally accepted. Heitler, for one, thought it was "totally wrong"

---

[15] Source in Portuguese (translation mine):

Voltando para a Italia, depois de uma outra publicaçao, eu disse que podese formular o grupo de Lorentz no espaço de momentos em energia. E quando temos momento e energia, pela indeterminaçao náo sabemos nada de espaço e tempo. De forma que nem o problema do ponto pode ser discutido.

([Wat75], pp. 6–7); Heisenberg, too, disliked it, and considered it rather artificial.[16] Despite this, 11 years later Feynman ([Fey48b], p. 1438) used Wataghin's cutoff in his famous relativistic regularization of QED, and credited him for it.

Relativistic causal order

By now it is clear that one of the most serious obstacles that threatens the incorporation of fundamental length into field theory is the failure of relativistic causal order at that length scale. It may therefore prove instructive to pause and see exactly how this failure "sneaks in."

Mathematically, the problem starts by dealing with infinitely many degrees of freedom, where complete knowledge of the initial conditions in a confined region in space is nevertheless insufficient to secure existence and uniqueness of solutions. This point was expressed succinctly in a rarely cited paper of von Neumann, where the example was a box with low energy particles whose energy could fluctuate indefinitely if acted upon by high energy particles outside the box:

The situation is similar to that which arises in the Newtonian theory of gravitation. In that theory, although matter distributed uniformly over a spherical shell $(r, r + dr)$ will have no resultant attraction at the center, if fluctuations occur there will be a resultant attraction and in an infinite universe approximately uniformly filled with matter the effect of the distant masses will lead to divergent results. In the present electrodynamics the particles of high energy lead to an analogous divergence. One might say: the distant regions in momentum space have the same divergence-generating effect in (Maxwellian) electrodynamics as the distant regions of common space have in Newtonian gravitation theory. Thus in both theories the fundamental postulate which requires the existence of "closed systems" is violated.

([vN36], pp. 250–251)

From this perspective, the problem of "runaway solutions" – the one that Dirac encountered in his classical theory of the electron, and which he solved with the extended electron – is conceptually equivalent to the problem of self-energy divergence Wataghin set out to solve with his cutoff of high frequencies. Both were trying to tame singularities, or failure of existence and uniqueness of solutions of the equations of motion. In the classical case, Dirac noticed that the singularity could be removed if the particle could be thought of not as a point but as an extended sphere. The anomaly of "runaway solutions" is then remedied, as the force acting on the particle need not traverse the distance to its center of mass in order for it to accelerate. But the price Dirac paid for taming the singularity was action-at-a-distance inside the sphere: the force acted on every point on the surface of the sphere, and in particular on points separated by the radius r_0, simultaneously, their space-like separation notwithstanding.

---

[16] In a letter to Pauli from 1937, quoted in [Kra95], p. 408, he offers the following harsh verdict: "Wataghin's work is not worth much; this Wataghin cannot think clearly."

This means that "closing" the system by restricting its momentum space, as von Neumann suggests (and Wataghin implements via his cutoff), entails action-at-a-distance below a certain length scale in position space. Agreed, as Dirac and Wataghin note, the same cutoff ensures that there is no way to resolve distances smaller than the postulated fundamental length, or, equivalently, to entertain inter-actions at higher energies, and yet, this means that relativistic causal order, or the temporal precedence of causes over their effects for all observers, ceases to be applicable at these scales.

Wataghin's notion of non-local field theory was first developed in the classical domain, where it fueled the revived interest in the extended electron, starting with the work of Bopp [Bop40], who suggested extending the notion of a charge density of a relativistic point charge $\rho(\mathbf{r})$, commonly expressed by the four-dimensional δ-function $\rho(\mathbf{r}) = e\delta_4(\mathbf{r} - \mathbf{z})$ where $\mathbf{z}$ is the position of the charge, by replacing its covariant form with the function $f(x)$, the *form factor*, that characterizes the shape of the charge, and that has all the same transformation and normalization properties as the δ-function:

$$\delta_4(x - z) \rightarrow f(x - z). \tag{4.19}$$

Classical non-local field theories dissolve the self-energy problem, and yield a stable electron (provided suitable terms, also known as Poincaré "cohesive forces," are added, that compensate for the Coulomb repulsion), and, with a suitable choice of the form factor, can also eliminate the runaway solutions ([McM48], p. 334). And yet they have the same non-local features acknowledged by Dirac, namely, the interaction term in the Lagrangian is replaced by a double integral that depends not on the product of the potentials and the charge density at the same spacetime point, but on the product of the two quantities at different points, with a weight factor which, to be relativistically invariant, must be a function only of the invariant distance between the points. The weight factor is large only when the distance is small, say, of the order of r_0, and vanishes rapidly as this distance exceeds r_0. This means, again, that inside the electron, or, equivalently, at time scales smaller than $r_0/c \sim 10^{-23}$ seconds, relativistic causal order breaks down.

The developments of renormalization techniques by the end of the 1940s weak-ened to some extent the necessity of introducing fundamental length into field theories. Nevertheless, in the 1950s there was a revived interest in non-local quan-tum field theories,[17] encouraged by contributions from Nobel prize laureate Hideki Yukawa [Yuk50],[18] that continued in the 1960s mainly in the Soviet Union (see

---

[17] Wataghin himself contributed to this program in a series of papers he published in *Il Nuovo Cimento* between 1951 and 1963.

[18] The inspiration for Yukawa's interest in high momentum cutoffs was Born's [Bor38a, Bor49] "principle of reciprocity." We shall say more about this principle in the following chapters.

[Kir67] and references therein). It is safe to say that while this research program has succeeded in overcoming almost all the technical issues to some extent [Kir62, Kir64, LK65], the failure of relativistic causal order remains unsolved [SW54, Mar73]. As Pais and Uhlenbeck, in their thorough analysis of non-local field theories, put it:

All in all, it seems to us, that the most hopeful approach which the present investigation possibly indicates is one, where, . . . , the possibility of an ordering of space–time events is no longer a strict requirement. But it should again be emphasized that here we have come to the limits of the foundations of our present picture of the physical world. An attempt at loosening up these foundations by relinquishing the "causality in the small" might throw more light on the situation.

([PU50], pp. 164–165)

Clearly, for non-local field theories to be taken seriously, further conditions should be imposed on the form factor such that the abnormal a-causal effects in the microscopic domain, for example, an electron starts accelerating before the force exerted acts on it, would remain unobserved in the macroscopic domain (in agreement with our experience). In this respect it was shown [CP53] that failure of relativistic causal order could remain macroscopically unobserved if the form factor in momentum space were smooth and had several bounded derivatives. In particular, large-scale causality is ensured if derivatives of arbitrary order exist.

Unfortunately, further investigations [PU50, Pau53] demonstrated that even when these restrictions are imposed on the form factor, when additional conditions are met so that the action of the non-local field theory obeys gauge invariance [CP54] and unitarity [Mar74] (as well as the general requirement that the non-local theory reduces to the local theory of distances larger than r_0), the divergences – which were the main motivation for these theories – cannot be removed, and while proponents of non-local field theories do find some loopholes in these no-go proofs (see, e.g., [Lez66]), the only way out seems to be to obtain an effective cutoff for the interaction with time-like rather than with space-like field components [Fey48a]. However, this would destroy the correspondence with the classical field equations. For example, to first order an electron would not be influenced by any static field.

Non-locality

Since violations of relativistic causal order seem to be the thorniest drawback of these theories, one may ask how serious a drawback is it? After all, to many a layman, non-relativistic QM also admits non-locality, and some interpretations of non-relativistic QM, such as Bohmian mechanics or the GRW theory, are also non-local. Do these theories violate relativistic causal order too?

The point, of course, is that there are different notions of non-locality at play here. To start with, in QM, non-local correlations exist between space-like separated events, in the sense that no local hidden-variables theory can reproduce them; this is the essence of J. S. Bell's famous theorem [Bel64]. This means that while the outcomes of space-like separated experiments are correlated (in the philosophical literature this feature is called "outcome dependence" [Jar89]), interactions in the theory (say, between a measuring apparatus and the observed system) are always local, and causal propagation (sometimes called signaling) between the sides of the experiment is always subluminal. In other words, relativistic causal order is maintained in orthodox QM, its non-local character notwithstanding (for a famous no-signaling theorem in QM see [GRW80]).

To see this, it is sufficient to note that in QM the empirical content of theory, namely the statistics of one experiment, does not depend in any way on the question of whether or not the other (space-like separated) experiment was done; as the no-signaling theorem shows, the reduced density matrix is the same in both cases. In Bohm's theory, on the other hand, the non-locality is of the latter, more serious, kind, namely, it actually makes a difference to the result of one of the experiments whether the other, space-like separated experiment, was performed or not, which means that in Bohm's theory interactions, and not only outcomes of the experiments, are non-local (in the philosophical literature this feature is called "parameter dependence" [Jar89], for a detailed and illuminating analysis see [Alb92], Ch. 7).

Bohm's theory allows superluminal signaling and therefore violates relativistic causal order. From a spacetime perspective, this feature amounts to introducing a preferred foliation to spacetime structure [Mau96]. Bohm's theory, however, incorporates a "mechanism" that keeps the preferred foliation unobserved [GDZ92]: the difference in predictions between QM and Bohmian mechanics would be evident (and the preferred foliation be detectable) only if one could achieve actual infinite precision in measurements. Lacking an ability of this sort, there is no way to detect superluminal effects in Bohm's theory *in practice*, although, *in principle*, the theory admits a preferred and absolute – albeit undetectable – frame of reference, in contrast to Einstein's special theory of relativity, where no such frame is supposed to exist.

The difference between Bohmian mechanics and QM is thus "a difference in philosophy" ([Bel87], p. 77). Yet the theory has so far resisted any generalization to the relativistic domain. Which brings us back to the case at hand of non-local field theories, whose non-locality, although a non-locality of interactions, still differs in two important ways from the non-locality of Bohmian mechanics.

First, in contrast to Bohmian mechanics, these theories are formally relativistic, as they are constructed not as an alternative to classical electromagnetism, but as more fundamental theories, that reduce to (a non-linear version of) the latter at macroscopic scales. Second, while they, too, purport to incorporate in their formalism a "mechanism" that keeps the microscopic non-local effects unobserved,[19] the inability to falsify these theories is regarded as only temporary: a matter of *practical* resolution capabilities, and not a matter of *principle*. A finer resolution, applicable at shorter length scales, could give rise to anomalous predictions. Consequently, since one need not resort to the "forever-unattainable-actual-infinite-precision" to test such theories, the difference between them and standard field theories is not only "a difference in philosophy," but also an empirically meaningful difference. In this regard, the more appropriate analogy here is between non-local field theories and the research program of subquantum theories (e.g., ['tH97]). These are intended, as in the case of the former, to give the same predictions in the macroscopic regime, but different predictions at higher energy scales, in their case, the Planck scale.

The methodological lesson here is that the introduction of fundamental length into a field theory must have, on the one hand, *in principle* detectable consequences in domains yet to be explored, but, on the other hand, must also be in accord with current experimental data. It is thus clear why physicists were reluctant to pursue theories that incorporated fundamental length both in the context of classical electromagnetism and in the context of QED. Since one of the most important consequences of this notion is the failure of relativistic causal order, and since contemporary experimental data indicated no such feature, there was no justified empirical motivation for its introduction. The only motivation was methodological, as a remedy for theoretical lacunae such as the divergence problems that saturated these theories. Once a conservative solution to these problems was found, namely, renormalization, there was no need to endorse more radical alternatives.

The remaining challenge to non-local field theories, however, stems from results such as [CP54, SW54, Mar74], which demonstrated that there is no apparent way to reconcile these theories with other formal conditions they had to satisfy, for example, gauge invariance and unitarity, while keeping their original motivation intact. These results did not show that no other way *could* exist; they only applied to contemporary methods.[20] It thus remains a challenge – maybe the most important challenge – that proponents of the notion of fundamental length face, namely, to come up with an empirically meaningful theoretical construction that includes a

---

[19] Recall that superluminal signaling becomes apparent only when one is able to resolve wavelength at the order of the fundamental length (both in the classical case and in the case of QED, this length is of the order of the radius of the electron, namely, 10^{-13} cm); for all practical purposes, at scales larger than this, interactions can be thought of as occurring at a dimensionless point, as Born and Infeld, as well as Wataghin and Dirac, have emphasized.

[20] For a construction of, e.g., a non-local gauge invariant electrodynamics that is relativistic invariant and unitary, see [Kir67].

fundamental length whose breakdown of relativistic causal order "in the small" can be "washed out" in the macroscopic regime, while satisfying additional well-confirmed mathematical requirements such as unitarity and Lorentz invariance. We shall inquire whether a construction of this sort is possible in Chapter 8.

Before we move on, let us mention another attempt at introducing fundamental length into the theory of matter, that went practically unnoticed in the 1930s.

Arthur March

Arthur March was an intriguing figure, a theoretical physicist and a talented writer [Kra95]. During the 1930s and over almost two decades March developed an extreme relationalist-operationalist philosophy, inspired by a casual remark of his colleague, mentor, idol, and sometimes academic rival, Erwin Schrödinger [Sch34, Kra92b]. We shall discuss March's ideas on fundamental length in the next chapter, as they are part of the prehistory of quantum gravity. However, in 1951, in a book on quantum theory of fields published in English, March devoted a chapter to a heavily truncated version of his ideas on fundamental length, applied now to the theory of the interaction between matter and radiation, where they serve, again, as a means for eliminating divergences.

March ([Mar51], Ch. 10) divides QED divergences into two classes. The first has its origin in the classical notion of a point charge. The second, which he describes as germane to quantum theory with no classical counterpart – although in ([Mar51], p. 230) he admits, as von Neumann does [vN36], that these divergences, too, result from the notion of a point-coupling,[21] albeit with a connection less evident than in the case of the electron's self-energy – includes all these cases where the theory is forced to admit waves of an arbitrary small wavelength, for example, vacuum polarization, the zero-point energy, or the magnetic moment.

Given the source of the second class of divergences, March believed that the only remedy for them was the introduction of a principle that limits the number of waves by removing the effectiveness of those waves with a frequency exceeding a certain limit. A naive application of this principle, however, would conflict with relativistic invariance, hence the bound on wavelengths must be frame independent. What could serve as an appropriate principle is the introduction of a new universal constant l_0, that, analogously to $\hbar$ and c, limits *in principle* the ability of observation, to the extent that one cannot ascertain the position of a particle at rest with an accuracy greater than the error l_0. The fundamental length enters, again, as a limit on spatial resolution.

---

[21] Recall that according to von Neumann, the case of infinite degrees of freedom in a box, mentioned by March, is analogous to loss of existence and uniqueness of solutions in position space due to (Newtonian) gravitational action-at-a-distance.

The fundamental length l_0 is then introduced relativistically into the interaction Hamiltonian between the field and the particle as the minimal radius of a sphere, below which no interaction takes place. The alteration of the interaction Hamiltonian occurs in such a way that momentum is always conserved. Consequently, the corresponding matrix element for, for example, the absorption of a photon by an electron which is initially at rest for wavelength $\lambda \gg l_0$ remains unchanged, but for $\lambda \ll l_0$ it is reduced to one-half the classical value. For emission the result is that an electron at rest is incapable of transmitting photons with wavelength smaller than l_0.

From these corrections to the interaction Hamiltonian, March ([Mar51], p. 277) concludes that it is impossible to distinguish the positions of two particles by means of a diffraction experiment performed with light rays, if the distance between them is less than l_0. In such an experiment, radiation would be required with wavelength of the order of magnitude corresponding to the distance to be measured. The zero-point energy divergence is now eliminated because waves with $\lambda \ll l_0$ cannot be reflected from the walls of the cavity in which the radiation is enclosed, as neither free, nor fixed, electrons are able to deflect a photon by a finite angle.

March's theory resembles Wataghin's theory in its strategy and its details (for these see [Kra95], pp. 419–433 and references therein), but it differs from the latter in two interesting aspects. First, while the immediate motivation for Wataghin's crude cutoff was a mismatch of experiment with theory in the context of *Bremsstrahlung* ("stopping"), March's theory was motivated by purely theoretical considerations. Whereas Wataghin's ideas would seem unmotivated after the discovery of the meson, March could show that the phenomenon of *Bremsstrahlung* is independent of l_0. More generally, for most interactions March's l_0 is practically insignificant, its role becoming apparent only in the elimination of divergences.

On the other hand, March believed that his fundamental length became extremely important for quantum field theories of higher energies than QED, for example, meson theory, and that the overlap between the electron radius and the length scale at which the nuclear force becomes dominant – an overlap that thus far had seemed a contingent coincidence – was an inevitable consequence of l_0 ([Mar51], p. 284).

The microscopic non-local effects that come with the fundamental length bothered March very little, as, according to his relationalist-operationalist philosophy, space is nothing but the relation between atomic bodies, and since the theory of matter he developed deems distances shorter than l_0 unobservable *in principle*, no meaningful discussion can ensue on physical behavior at scales shorter than this. In this sense, he saw his theory, describing as it did only the space of our finite measurement results, as completely compatible with the spacetime continuum [Mar36]. This is yet more evidence that many working physicists at that time derived no ontological consequence from the inability to resolve distances shorter than a certain length. As shall become clear towards the end of the book, without phenomenological consequences for spatial discreteness, namely, observed deviations from the well-established empirical principles that characterize field theories, the decision to endow the limit on spatial resolution with ontological weight is not forced on us, and is purely a matter of taste.

March's ideas were practically ignored by the physics community.[22] Part of the reason for this had to do with March's own ignorance of contemporary particle physics, for example, arguing, as he was ([Mar51], p. 289), that his theory could remove an anomaly in particle physics that was already known to be non-existent. But the relationalist-operationalist philosophy that fueled his ideas is interesting (although it, too, was ignored by philosophers, past and present), and we shall pursue it further in the next chapter.

4.4 The rise of renormalization

While non-local field theories (and extended particle theories in the classical domain) adopt a finitist view, rejecting as they do the actual infinity of a dimensionless point-particle, by the end of the 1940s the majority of the physics community had settled for a third option, namely, endorsing potential infinity and the mathematical method of renormalization. This option has so far proved extremely efficient for all known physical interactions except gravity, and will be discussed next.

4.4.1 The floating cutoff

As in the case of the prehistory of QED, also here, the story of the renormalization program is well known, and includes many contributions from different sources (see, e.g., [Wei77, Ara89a, Sch94, CS93, Cao97]). As our purpose here is only to elucidate the conceptual differences or similarities between the renormalization approach and the approaches discussed above, we shall only briefly gloss over the major milestones, emphasizing those that exemplify best these differences or similarities.

We mentioned earlier the source of the ultraviolet divergences in quantum field theory: there exists an infinite number of states for the electron, and an external field can cause a transition between any of them [vN36]. We also noted that – while this situation can yield divergences with no classical counterpart – in essence it is just another manifestation of the singularity of the point-coupling, already present in classical field theory. In QED the point-coupling entails an unrealistic

---

[22] Heisenberg was an exception ([Hei62], p. 6, 32), although he, too, rejected March's use of modern physics. This support seems even more awkward given Heisenberg's dislike of Wataghin's cutoff, which is identical to March's.

contribution from virtual quanta – the producers of force – with high energy and momentum. The most natural and straightforward way to deal with divergences of this sort was discussed above, namely, one can introduce a cutoff to the integral, and impose an upper bound on possible energy levels of the interactions that the theory is able to describe. This move is tantamount to introducing a characteristic length scale, below which the theory ceases to be applicable. As we have seen, it is hard to reconcile this cutoff with additional constraints that QED is required to satisfy.

An alternative method, developed throughout the 1930s and 1940s, was to identify the infinite terms in the perturbation series of the interaction, and to separate them from the finite ones, so that they could be subtracted in some way or another from the formalism. This was done for both the mass and charge of the electron (see [Cao97], pp. 185–203). The method matured into two approaches. One, developed by Kramers, Bethe, Lewis, Schwinger, and Tomanaga, identified directly the finite terms with the observable quantities and the divergent terms with their corrections, and removed the latter from the expressions describing real process by redefining the mass and charge. This redefinition was called *renormalization*; it abstained from any reference to very high energy processes and the related small distance and inner structure of the electron. The justification for this procedure would probably sound familiar to the reader:

QED unquestionably requires revision at ultra-high energies, but is presumably accurate at moderate relativistic energies. It would be desirable, therefore, to isolate those aspects of the current theory that essentially involve high energies, and are subject to modification by a more satisfactory theory, from aspects that involve only moderate energy and are thus relatively trustworthy.

([Sch48b], p. 416)

The other approach, attributed to Feynman [Fey48b], and Pauli and Villars [PV49], was called *regularization*. In contrast to its counterpart, it made explicit use of a cutoff, but this magnitude, say ϵ, also called the *regulator*, was allowed to be arbitrarily small. The correct physical result is obtained in the limit in which the regulator vanishes ($\epsilon \to 0$), but the virtue of the regulator is that for any of its finite values, the result is finite. However, the result usually includes terms proportional to expressions like $1/\epsilon$ which are not well defined in the limit $\epsilon \to 0$.

It was Dyson [Dys49] who synthesized the two approaches: when regularization, the first step towards obtaining a completely finite and meaningful result, is followed by the related, but independent, technique of renormalization (which identified the physical quantities – expressed by seemingly divergent expressions such as $1/\epsilon$ – as equal to the observed values), one can calculate a finite value for many other quantities that looked divergent (for a clear exposition of this procedure see, e.g., [Del04]). In particular, if after the redefinition of mass and charge, other processes are insensitive to the value of the cutoff, then a renormalized theory can be defined by letting the cutoff go to infinity. A theory is thus called *renormalizable* if a finite number of parameters (such as mass and charge) are sufficient to define it as a renormalized one.

The ideas of mass and charge renormalization, implemented through a meticulous analysis of the symmetry properties of QED, namely, its Lorentz invariance and gauge invariance, made it possible to formulate algorithmic rules to eliminate all the ultraviolet divergences that had plagued the theory. The success of renormalization theory in accounting for the Lamb shift, the anomalous magnetic moment of the electron and of the muon, and the radiative corrections to the scattering of photons by electrons, to pair production, and to *Bremsstrahlung*, was spectacular [Sch02], with a fabulous precision that is rarely found in any other domain of theoretical physics.

The predictive success clearly contributed to the credibility of the renormalization procedure, to the extent that some physicists such as Bethe and Schwinger even regarded it as a guiding principle, a heuristic for model building and theory construction ([CS93], pp. 43, 49). This belief was fully borne out: the procedure also proved to be important in formulating quantum chromodynamics, the quantum field theory for quarks, and electro-weak theory ([Wei80], pp. 517–520).

A minority of physicists, however, were not swayed. Perhaps the most vocal critic of the renormalization program was Dirac who, ironically, was a major player in its prehistory. According to Dirac ([Dir63], pp. 49–50), the predictive success of the program might as well have been "a fluke," and should certainly not be considered as a guide to its truth. He held his skeptic position for the rest of his life, and voiced it on many occasions, arguing that the renormalization method is inherently inconsistent (see, e.g., [Dir81]).

Presumably what bothered Dirac was the arbitrary neglect of infinitely *large* terms, which was contrary to the standard mathematical practice of neglecting only infinitely *small* terms ([Dir78], p. 36). The problem can also be stated as the following dilemma ([Roh07], pp. 128–129): either the renormalization method is dependent on the cutoff ϵ, or it is not. If it is dependent, then $\epsilon \to 0$ does not make sense; if it is not, then why choose a cutoff ϵ to begin with?

In response, supporters of the renormalization program argue ([CS93], pp. 69–71) that the inconsistency lies with the skeptics, who are committed to an implicit speculation about the inner structure of the relevant elementary particle (in the case of QED, the electron), which is sensitive to the details of the dynamical processes at high energy. Mathematically, this assumption is manifest in the divergent integrals. Metaphysically, the assumption implies that there exist *more* elementary constituents of the physical particles described by the fields appearing in a

Lagrangian, and this contradicts the elementary status of the particles, or that of the fields, as basic building blocks of the world.

Another line of defense is methodological: the renormalization program adopts a notion of potential infinity, i.e., particles are treated as quasi-points. This model seems to be the appropriate mathematical device needed in a transition period, when one has no way to probe the high energy regime. Now, a hardcore reductionist might hold that this construction is only temporary, and should be discarded once the most fundamental theory, a "theory of everything," is discovered; but regardless of whether a theory of this sort even exists, in the mean time, to make progress, even the reductionist would need a seemingly point-particle model to make sense of its inner structure, once enough energy was available. For example, the analysis of deep inelastic lepton–hadron scattering would be impossible if there were no analysis of elastic scattering of the point-like partons as a basis ([Cao97], p. 206). Thus from this methodological perspective, the renormalization program is actually a necessary precondition for quantum field theories.

This methodological defense was substantiated in the late 1960s and the early 1970s, when it became clear that (1) the existence of a limit as $\epsilon \to 0$ and (2) the independence of the final result from the regulator, are both highly non-trivial facts, for which the underlying reason is the existence of second order phase transitions.

The condensed matter technique that describes the latter, known as the renormalization group (RG henceforth), was developed by Leo Kadanoff [Kad66] and Kenneth Wilson [WK74, Wil75].[23] With this approach, QED and other quantum field theories could now be regarded as only approximating an ideal correct theory. The approximation breaks down at high energies, and so one must discard the high energy sector of the momentum space. But due to the non-trivial points listed above, one can now restrict attention to the finite quantities that will be accurate no matter what the cutoff is, as long as it is large enough that the correction terms above it differ negligibly from their limiting values. Moreover, even if, sometimes, taking the limit as ϵ goes to zero is not possible, if the regulator only gives reasonable results for $\epsilon \gg 1/\Lambda$ (because of some new physics below Λ), and one is working with scales of the order of $1/\Lambda'$, regulators with $1/\Lambda \ll \epsilon \ll 1/\Lambda'$ still give pretty accurate approximations.

One approach that the RG program has spawned is now called "effective field theory" (EFT henceforth) [Geo93, Bur07]. Scale-dependent theories are developed that are applicable only up to some energy scale, beyond which new physics is required. QED, for example, is intended to describe electrons and photons and the

---

[23] The seeds of the renormalization group approach can be found already in the early 1950s, see [Ara89b] and references therein.

interactions between them, and nothing more. The applicability scale of the theory is thus 10^{-13} cm, and it virtually ignores whatever might be going on between heavier and heavier particles whose characteristic lengths are shorter than this scale.

What gives legitimacy to this maneuver are two facts [AC75]: first, the production of heavier particles requires more energy, and this energy is unavailable at that scale; second, lighter particles interact relatively weakly, so for all practical purposes we can ignore both.[24] When we extend QED to a theory whose energy scale is higher, we incorporate into that theory particles whose characteristic length is shorter, and refer to them explicitly in the formalism. In this way we arrive piecemeal at more accurate theories without having to entertain energy scales outside our current theory.

Clearly such an anti-reductionism is not forced on us;[25] the EFT approach is not the only method consistent with the RG view. Using the notions of "Gaussian fixed point" [WK74] and "asymptotic safety" [Wei78], one can show that a renormalized quantum field theory with a finite but unbounded cutoff can capture exactly the same physics as a continuum theory of fields with point-like particles ([HW95], p. 183). In other words, because of the non-trivial facts (1) and (2) above, the RG program succeeds in describing physics of the limit; a renormalized theory is thus scale invariant, and can be regarded as an approximation to a theory with an unrestricted domain of applicability, that holds at all energies. Similarly, nothing in the renormalization program forces any ontological commitment to spatial discreteness: as in the case of non-local field theories, the cutoff here does not necessarily *entail* that spacetime has a lattice structure. Even in the case of QED, which is not asymptotically safe, one can still interpret the theory as a small lattice approximation (and, given its predictive success, a very good one at that) of a more general continuum theory.

Moreover, by looking at different renormalization schemes that practicing physicists actually employ, one can identify two types of empirically equivalent EFTs – Wilsonian EFTs and continuum EFTs [Geo93]. These are non-trivial examples of empirically equivalent theories in so far as, in the context of a given high energy theory, they make the same low energy predictions, but they suggest different ontologies. Continuum EFTs support an ontology that endorses a continuous spacetime, whereas Wilsonian EFTs require space to be discrete and finite. Consequently, the EFT approach too, does not necessarily entail that the lattice structure,

---

[24] Actually, this is not quite right, as we also need spontaneous symmetry breaking to get from the electro-weak theory to QED. In this sense Fermi's theory of the weak force is a neater example, see [HW95], p. 186.

[25] On the debate over the preferred methodology in high energy physics see, e.g., [And72, Wei87, Cas02].

borne out by the cutoff, should carry any ontological weight; one can still hold the view of the continuum.[26]

The upshot is that the renormalization program is a non-committal compromise between two extremes, a kind of a "third way": on the one hand continuity and actual infinity (of the point-particle) and, on the other hand, discreteness and the rejection of point-coupling (with the introduction of a fundamental length and a lattice structure to spacetime). Given that none of the empirical evidence in the low energy sector distinguishes between these two extremes, one is left in the same situation we have depicted in Chapter 2, namely, with two consistent mathematical structures that are equally applicable in the physical world. This feature of the renormalization program is evident not only in the discussion on the ontological status of spatial discreteness, but also in debates on the methodology of particle physics, as the program leads to a compromise between continuum theories which are too phenomenological, and lattice theories which are too far removed from available experimental data, whereas current empirical evidence equally supports both theory types.

4.5 Philosophical ramifications

We began this chapter by asking what were the motivations for introducing fundamental length to theoretical physics in the 1930s, and what role did it play. We are now finally in a position to answer these questions.

4.5.1 Methodology

First, from a methodological perspective, the notion of fundamental length was introduced as a possible solution that could mitigate the anomalies between the newly born theory of QED and the empirical data that were accumulating fast during the 1930s. With the clarity of hindsight, this motivation can be seen today to be completely unfounded, but one should recall that in the 1930s the stream of empirical data overwhelmed the theoreticians, growing faster than the theoretical developments. Phenomena such as *Bremsstrahlung*, cosmic showers, explosions, and pair creation were taken by, for example, Heisenberg and Wataghin, to motivate a high momentum cutoff, but, in both cases, later theoretical developments made this notion redundant, at least in the context it was introduced. Moreover, the contingent matter of fact that the electron radius $r_0 \approx 10^{-13}$ cm is also the scale at which nuclear phenomena become relevant contributed to the theoretical confusion

---

[26] For an elegant analysis of the confusion that exists when one carelessly underspecifies which EFT version of quantum field theory one endorses, namely, the continuum EFT (e.g., [Cas02]) or the Wilsonian, lattice EFT (e.g., [CS93, Wal06, Fra09]), see [Bai11].

in the 1930s, and to the idea that QED breaks down at length scales shorter than r_0. This also explains the tendency to interpret the said cutoff as a limit to the applicability of the theory, and not as a feature of spacetime.

Next, from a theoretical perspective, a fundamental length was a natural, almost instinctive, candidate for a solution of the ultraviolet divergences in field theory. The difficulties in reconciling it, however, with further theoretical requirements such as Lorentz invariance, gauge invariance, unitarity, conservation laws, and, above all, relativistic causal order, persisted throughout the two decades that preceded the birth of renormalization theory. This became the most serious obstacle to its credibility, to the extent that once a solution such as renormalization was in the offing, which required minimal deviation from well-accepted (and well-confirmed) relativistic notions and symmetry principles, the idea of fundamental length, manifest as an actual cutoff in momentum space, and as a limit to physical interactions, was almost abandoned in mainstream physics.

Not everybody was convinced, however, and the rationale behind this choice of theory is worth elaborating on.

For most wave fields propagating in a medium (e.g., sound waves or vibrations of strings), the description of the field functions as continuous functions of continuous parameters $\mathbf{x}$ and t, as well as the description of changes in the fields at a point $\mathbf{x}$ as determined by properties of the fields infinitesimally close to $\mathbf{x}$, are considered an idealization which is valid for distances greater than the characteristic length that measures the granularity of the medium. For distances smaller than that, the theories must be modified. The electromagnetic field, however, is a notable exception, and this fact was only acknowledged after Einstein's special theory of relativity removed the necessity for an underlying medium for light propagation.

The assumption that the electromagnetic field description is correct in an arbitrarily small region of spacetime was challenged, however, by the ultraviolet divergences, and while renormalization theory has sidestepped these difficulties, many have felt that the said divergences are symptoms of yet to be discovered new physics.

We have already mentioned Dirac's dislike of the renormalization approach. Another key figure in the development of the program, Schwinger, was no less critical:

But, we may ask, is there a fatal fault in the structure of field theory? Could it not be that the divergences – apparent symptoms of malignancy – are only spurious byproducts of an invalid expansion in powers of the coupling constant, and that renormalization, which can change no physical implication of the theory, simply rectifies this mathematical error? This hope disappears on recognizing that the observational basis of quantum electrodynamics is self-contradictory ...

([Sch58], p. xv)

Schwinger then explains this harsh verdict: virtual quanta – the dynamical variables in the theory – are treated as free particles, uninfluenced by their coupling to the electromagnetic field, but there is no way to attribute this status to these particles as they can never be observed as such, exactly because their localization with arbitrary precision requires for its realization a coupling with the electromagnetic field, which can attain an arbitrary large magnitude. The divergences, thus, are not a mathematical artifact at all; instead, they deny the very observational basis of the theory!

He then concludes:

... a convergent theory cannot be formulated consistently within the framework of present spacetime concepts. To limit the magnitude of interactions while retaining the customary coordinate description is contradictory, since no mechanism is provided for precisely localized measurements.

([Sch58], p. xvi)

These remarks warrant a short detour into the debate on the status of virtual quanta, a debate which exemplifies the clash between the two world pictures, the field and the particle, and illustrates once more the recurrence of the arguments from math, surveyed in Chapter 2, for and against the viability of the description of physical space as continuous or discrete.

The common view among philosophers of physics is that virtual quanta have no physical significance (e.g., [Red88], p. 20; [Wei88], p. 46; [Tel95], p. 139). Appearing as they do in Feynman's diagrams, which are ultimately just a tool for calculation, virtual quanta should be interpreted, or so the story goes, as merely playing a formal role. This opinion is enforced by the idea that there are no convincing positive arguments for a realist interpretation thereof, while there are at least two arguments against this.

One argument against a realistic interpretation of virtual quanta is an argument by analogy ([Fox08], p. 35). As Redhead ([Red88], p. 20) puts it, "To invest them with physical significance is like asking whether the harmonics really exist on the violin string?" The idea here is that as much as we do not believe that a string is vibrating with all the frequencies that appear in the weighted sum of its normal modes, but only with one frequency, namely the said weighted sum, we should not endow the virtual quanta which appear in the mathematical expansion of the S-Matrix with ontological weight. The point is analogous to the idea that the empirical content of quantum (field) theory resides in the probabilities (cross-sections), and not in the amplitudes which the superposition the state is in consists of. These amplitudes are necessary for calculation of the probability (cross-section) but they are not real in any sense.

Another argument against the reality of virtual quanta is that it forfeits the classical intuition about a physical process unfolding in time, for it requires us to abandon the temporal character of the interaction in the scattering process: the Feynman diagram is seen as representing the overall, global, spacetime description of the scattering process, a black box without spatial parts, and more importantly, without temporal order. The idea here [Val11] is that if one tries to analyze the interaction in a scattering experiment from a submicroscopic point of view in an attempt at a temporal description of the scattering process, then there would be no way physically to describe the propagator: an "observer" who tried to determine the scattering amplitude of the virtual quanta may be able to identify a past (emission) and a future (absorption), but another, moving, observer may describe these in reverse.

I personally find the two arguments above unconvincing. First, similar to the amplitudes in the superposition, or the normal modes of the violin string, the virtual quanta still have a *collective* kind of existence (this fact is also acknowledged by the proponents of the said argument, see, e.g., [Fal07], p. 237): together they contribute to a scattering amplitude or transition probability, and so together they cause a real collective effect. This point may lead us to remote philosophical terrains: is a termite colony real, or do only individual termites exist? Physicists might shrug, and would probably say that there is no way, *in principle*, to observe virtual quanta, but this also is unconvincing: we cannot observe free quarks, yet we still believe they exist.

With respect to the second argument, as it stands, it strikes me as circular. True enough, relativistic causal order is so intuitive and such an essential part of our macroscopic experience, but this does not mean that it must hold at all possible scales. One cannot extend the intuition that works so well for the field concept at one scale, and use it as a premise in an argument against the denial of the same intuition at another scale, especially when (1) there is no evidence yet for the applicability of this intuition at all scales, and when (2) there is no argument that this intuition cannot be denied at one scale and hold at another.

The two arguments seem to me therefore at the very least incomplete: the first requires some additional story about collective versus individual existence; the second requires some additional story about the possibility (or rather the lack thereof) of violating relativistic causal order only below some designated scale, while maintaining it above that scale.

I suspect, however, that even after one completed these arguments, they would remain unconvincing, as they reflect the very type of argumentation we have encountered in our excursion into the philosophy of math. Take the first one. Its essential premise is that the basic building blocks which are inconsistent with the field concept cannot be physically individuated. But this premise rests on the notion of the continuum (the components in a superposition, or the modes in the vibration of the string, form an uncountable infinite, truncated only for practical purposes) which precludes such a physical individuation to begin with. Now take the second argument. It says that even if such basic building blocks could be individuated by some physical process, this process would revoke a very basic feature of spacetime that underlies the continuum, which again means that the continuum is assumed in advance.

That the price of maintaining a discrete picture (in this case, of particle exchange force) is the abandonment of our intuitions about features of spacetime at scales below a certain resolution should be, by now, no news to the attentive reader. The claim that this price is unaffordable, however, cannot be defended *a priori*. What I find ironic is that the metaphor of exchange forces that led to Feynman diagrams, and that had its roots in the intuitive picture of chemical reactions unfolding in spacetime [Car96], has ended up in a denial of this intuitive spatiotemporal picture. And yet, ironic as it may be, proponents of the continuum cannot use the denial of this intuitive picture against their opponents without begging the question.

The relevant lesson to our discussion from all this, I think, is that the debate on the status of virtual quanta is yet another facet of the debate on the applicability to physical space of the two world views, the continuum and the discrete. We already know from Chapters 2 and 3 that this debate cannot be resolved with philosophical arguments alone, and requires an empirical solution. For this reason it seems inappropriate to attack the applicability of the continuum perspective from a discrete perspective, as it is inappropriate to attack the applicability of the discrete perspective from the continuum perspective.

What Schwinger's criticism does reveal, however, is how strong is the pull of the particle picture. While physicists always work with quantum fields, they almost invariably describe the physics using particle language: the force between two charges, for example, is described as the exchange of virtual photons, when all calculations show that it is due to the Coulomb field of one particle on the other. Some physicists see this as a drawback and argue, as Bill Unruh does, that taking Feynman diagrams (and the virtual quanta they employ) seriously has "set back physics by 20 years, because it blinded physicists to what are called non-perturbative effects – effects which are often trivial when looked at from the field point of view" ([Unr10], p. 138). This, along with the price of the particle picture, may indicate why local field theories, whose properties can be described by continuous differential laws of wave propagation, are so ubiquitous in physics.

In their famous textbook on QFT from 1965, Bjorken and Drell ([BD65], pp. 3–5) address this question, first by citing the spectacular agreement of the predictions of QFT with experiments, and second, by simply admitting that no convincing alternative is in the offing. They argue further that, if one chooses to hold to the two essential principles of both non-relativistic QM and special relativity, namely, the method of quantization that requires a Hamiltonian, and Lorentz invariance, respectively, then the continuum follows *a fortiori*: the Hamiltonian – via Schrödinger's equation – requires continuity in time, and Lorentz invariance requires continuity in space.

This claim, as we have seen, is not quite right. Alternatives do exist, some of which even include a Hamiltonian (hence a continuity in time) and are (at least macroscopically) Lorentz invariant. What seems to be the deal breaker, however, is the notion of relativistic causal order. Spatial discreteness violates it and, according to Bjorken and Drell, while this may be resolved by retarded action (i.e., backward causation) between any two points of the lattice, this complicates the problem and yields no corresponding gain in understanding. From the empirical perspective, moreover, there is nothing but positive evidence for the applicability of special relativity in the high energy domain, and no concrete evidence for breakdown of relativistic causal order, and this justifies the current practice in physics and, moreover, constrains any attempt to deviate from this practice: any alternative to the continuum at length scales shorter than 10^{-13} cm should approximate it in that limit.

The lessons we have drawn from the discussion on the consistency and the applicability of spatial discreteness in the foundations of mathematics are finally manifest. There is nothing inconsistent in a field theory whose underlying spatial structure is a lattice. Actually, it is the other way around: the desire to establish finite predictions in a physical theory is in conflict with the notion of actual infinity, and so, as Schwinger notes, the very concept of a point event, which appears in the formulation of the condition of a local field, is incompatible with the measurability conditions in relativistic quantum mechanics. The impossibility of securing the condition of relativistic causal order in a non-local theory thus provides no physical argument against this theory. The only important thing is that there be no violations of relativistic causal order on a macroscopic scale, in accord with our experience.

Defenders of spatial discreteness, however, should proceed here with care. While abiding by the constraints imposed by current empirical evidence, they should still allow their theory *in principle* to be tested, so that it would yield different predictions in those regimes where relativistic causal order fails, or in domains yet to be explored. Otherwise, the question of spatial discreteness would remain purely metaphysical. Consequently, arguments that purport to establish that the non-local effects are hidden so deep that there is no hope of reaching that region experimentally [Efi72], or that the condition of relativistic causal order has no direct physical meaning [Kir67], are ultimately self-defeating. Here, as much as anywhere in physics, there is no victory in establishing that one cannot be proven wrong.

4.5.2 Interpretation

What about the role played by the fundamental length? Besides its ontological interpretation as a minimal distance that confers a lattice structure on space, it has also been taken to signify an epistemic limitation on measurement resolution. This limit plays a double role. First, it allows one to restore consistency to the mathematical description by eliminating singularities that arise from attempts at resolving actual infinities (e.g., the zero-point energy). Second, it delineates the limits of applicability of the mathematical description of the physical theory at hand.

It is important to emphasize that these two roles are in no way contradictory. That a certain theory is internally consistent does not make it universal in any sense: limits of validity can only be derived from empirical arguments, quite distinct from any logical analysis of internal consistency. The latter only serves to establish the theory's applicability in a given domain.

This distinction between the epistemic and the ontological interpretation of fundamental length reminds us once more of the difficulty in making the question of spatial discreteness an empirical one. Operationally, the role of characteristic length as an epistemic limitation on measurement resolution is similar in both QED and the classical theory of the electron: for all practical purposes, at scales of many orders above the electron radius, the electron can be idealized as a point-particle; at scales of the order of the electron radius, however, attempts at describing physics with the given theory are unsurprisingly unsuccessful.

The epistemic notion of characteristic length is thus relative to the theory at hand, and it indicates the scale below which the mathematical description breaks down. This is also the case in non-local field theories and in the renormalization program, where the inference from the limitation of resolution is only made to the applicability of the theory, and need not entail an underlying lattice structure.

The failure of the theory to resolve lengths shorter than the characteristic length may indicate, according to the ontological view, a breakdown in the essential features of spacetime structure (cf. Dirac's conclusion from the failure of locality below lengths of the order of the electron radius). In this case the characteristic length is also considered minimal. The epistemic view, on the other hand, is agnostic about the structure of spacetime, and regards this failure as a breakdown in the domain of applicability of the theory, a clue that additional physics and new forces are at play.

One may be tempted to identify the culprit in the inability to derive the structure of spacetime from the limitation on measurement resolution in the distinction between "characteristic" and "universal" length. On this view, that the spatiotemporal description of the field theory under discussion breaks down does not entail that the structure of the underlying spacetime on which the dynamics of the field theory unfolds is now discrete, only because the said length is not regarded as "universal," or "absolute," but just "characteristic," and only because the theory that incorporates it is not regarded as universal. The idea may be seen to be supported by the fact with which we opened this chapter, namely, that no universal notion of fundamental length appears so far in physics, although every elementary particle has its "own" characteristic wavelength below which the particle loses its identity. From this perspective, the decision to designate one of the characteristic lengths as "minimal" coincides with the decision to designate the respective theory as "universal." Since, without gravity, QED – or any other quantum field theory that does not include gravity, for that matter – cannot be considered universal, the characteristic lengths in these theories are thus in no way universal, and so there is no reason to draw any ontological conclusion from the epistemic constraint they represent.

But suppose we had a universal theory, superseding QFT and the Standard Model, and suppose further that this theory assumed, or predicted, a notion of fundamental length. Would this amount to a collapse of the said distinction between its characteristic and the universal nature? If the characteristic length were also universal, could it then play the dual role of epistemic bound on spatial resolutions *and* ontological spatial discreteness? Would this universality entail that the latter follows from the former?

I doubt that this will be the case. For an empiricist, the question of whether a theory is universal is, on final account, an empirical question and not a matter of methodological decision.[27] Even if one declared a theory that included a notion of minimal length as a limit on spatial resolution to be the final theory, proponents of the continuum could resist the said inference from this limit to the structure of space, as long as the said theory gave us no testable predictions for new phenomena, but only reproduced the known results of QFT and the Standard Model. In other words, if the "new physics" at the scale of the universal length turned out to be not new after all, there would be no way to resolve the issue (pun intended); both structures, the discrete and the continuous, could still be applied, no surprise, as appropriate descriptions of space.

---

[27] Although some physicists tend to forget that. See, e.g., the attempts to exorcise the Maxwell demon, or to generalize the second law of thermodynamics to black holes, or, more relevant to our discussion, the arguments for the necessity of quantizing gravity, discussed below.

The unification of quantum theory with the special theory of relativity has brought us back to square one, namely, to the comprehension that no amount of philosophical argumentation can persuade either of the opposing camps to change their views, and that only empirical evidence for spatial discreteness, i.e., a breakdown in the successful continuum based description, can settle the issue. In the next chapters we shall inquire what role was played by the fundamental length in the attempts to unify quantum theory with the general theory of relativity, and whether the challenge of empirical support for the hypothesis of spatial discreteness has finally been met.

5

Quantum gravity: prehistory

5.1 Outline

Our best theory of the constitution of matter, as we have just seen, cannot decide the question of spatial discreteness. Regardless of what its basic ontology consists of, whether particles or fields (for the debate on this particular issue see, e.g., [Fra08, Fra11] versus [Wal06, Wal11]), in both cases the spacetime structure that underlies this basic ontology can be regarded as either continuous or discrete. Current empirical evidence, moreover, while supporting a continuous structure down to length scales of 10^{-16} cm, is still consistent with a view that takes the theory to be a good continuum approximation of a more fundamental discrete structure. The justification for this view rests on two observations. First, quantum field theories have only succeeded in unifying all non-gravitational interactions; gravity – the last interaction to be unified – has so far been left out of the picture. Second, the physics 17 orders of magnitude below the contemporary observational threshold may be quite different.

In parallel with the developments in particle physics we surveyed in the preceding chapter, the first half of the last century saw many attempts at constructing a theory of discrete spacetime. Motivations for this program varied. Some were purely metaphysical (the notion of indivisibility of space), while others were methodological (unification of quantum theory with general relativity), or epistemological (a relational-operational limit on observability). While these motivations were mostly independent of the attempts – discussed in Chapter 4 – to introduce the notion of fundamental length into theories of matter, in some cases they were associated with the former program, and so, at least from a historical perspective, they were tied with the notion of the electron radius.

In this chapter we shall inquire about early attempts at discretizing spacetime.[1] We shall first present a historical account of these attempts, and then in the final

---

[1] The fascinating history of the discretization of time alone – leading to *chronons*, or time-atoms – will not be discussed here. The interested reader should consult, e.g., [KC94], pp. 442–449, and references therein.

section offer a philosophical analysis thereof. As we shall see, some of the conceptual problems we have encountered in the previous chapter will resurface and, once again, our aim will be to draw philosophical lessons from the attempts at reconciling them with well-established principles and confirmed empirical evidence.

In the final section we shall also address the "disturbance" view of the uncertainty principle, on which most of the intuitions and thought experiments rely, that underlie the reasoning behind the limitation on spatial measurement resolution, and in so doing offer an alternative to this view and some plausible, charitable, reading of the arguments associated with these intuitions and thought experiments.

5.2 Early steps

Einstein was one of the first to suggest that general relativity and quantum theory should be brought into a unified theoretical structure. As early as 1916, when discussing gravitational waves, he noted that one should expect gravitational radiation from the atom due to electron motion inside it, and that quantum theory should thus modify not only Maxwell's equations but also his newly formed theory of gravitation:

> Because of the intra-atomic movement of electrons, the atom must radiate not only electromagnetic but also gravitational energy, if only in minute amounts. Since, in reality, this cannot be the case in nature, then it appears that the quantum theory must modify not only Maxwell's electrodynamics but also the new theory of gravitation.
>
> *([Ein16], p. 696; translated in [Gor92])*

Einstein is of course known for his "rigid adherence" to the continuum of field theory, but a careful search in his letters and writings shows "another Einstein" who deliberated, sometimes even agonized, over the possibility of discrete space (see [Sta93] and Chapter 6). In a letter to Dällenbach from November 1916, for example, one finds the following remarkable insight:

> You have correctly grasped the drawback that the continuum brings. If the molecular view of matter is the correct (appropriate) one; i.e., if a part of the universe is to be represented by a finite number of moving points, then the continuum of the present theory contains too great a manifold of possibilities. I also believe that this "too great" is responsible for the fact that our present means of description miscarry with the quantum theory. The problem seems to me [to be] how one can formulate statements about the discontinuum without calling upon a continuum (space–time) as an aid; the latter should be banned from the theory as a supplementary construction not justified by the essence of the problem, [a construction] which corresponds to nothing "real". But we still lack the mathematical structure unfortunately. How much I have plagued myself in this way!
>
> *(Einstein in [Sta93], p. 280)*

That physicists in the 1920s did not appreciate the difficulties in achieving this goal is evident from the somewhat brisk remark by Pauli and Heisenberg that quantization of the gravitational field could be done easily, analogously to their method of quantizing the electromagnetic field, at least for the approximate case of weak fields, i.e., gravitational waves, where the field equations can be linearized (see [HP29], p. 32, and Section 4.3.1). The task, as is well known, turned out to be much more difficult then they assumed.

5.2.1 *Absolute uncertainty*

During most of the 1920s, the notion of quantized spacetime was mentioned anecdotally, but never made it beyond the fringe of theoretical physics. The context in which it did develop into a fruitful idea, however, was that of the minimal resolution of observation, which in the late 1920s and the beginning of the 1930s became related to discussions of the measurability of field quantities and the notion of absolute uncertainty.

Around 1930, almost all physicists believed that matter consisted of only two elementary particles, the electron and the proton ([KC94], p. 451). The notion of absolute uncertainty was suggested by Flint [Fli28] and Ruark [Rua28]. Both physicists included the electron mass in their thought experiments and drew the operationalist conclusion that, independently of Heisenberg's uncertainty relations, there exists an absolute uncertainty in position measurement. In later years this notion was developed further by Wataghin [Wat30a], and enjoyed support from authorities such as Pauli, Schrödinger, and de Broglie.

Ruark presented several thought experiments to motivate his idea. One of them included a measuring apparatus as small as the electron in size:

If the q's and p's of the measuring device, B, are of the same order as those of the measured object, A, then the changes in B caused by A must be considered. The coupled system $A + B$ must be treated as a whole, and the interpretation of a measurement consists in answering the question, what can we say about a coordinate of A when we have determined a coordinate of B with our gross instruments? An example is the measurement of the position of an electron initially at rest, by allowing it to scatter a light quantum of wavelength λ_0. If we attempt to measure the position very accurately by reducing λ_0, a limit is imposed by the increase in wave-length during the scattering process. For an electron, this increase is of the order h/mc, and this is the order of magnitude of the smallest length which can be resolved in such an experiment.

([Rua28], p 325)

The limit $\lambda = h/mc$ also fixes the minimal proper time interval to be of the size $\Delta t = h/mc^2$, where m is the mass of the electron. Flint, who attempted to unify electrodynamics, general relativity (in its Weyl version) and quantum mechanics, was led to the conclusion that the electron's worldline consists of discrete units of size h/mc, and that nothing of a smaller length is observable directly or indirectly in experiment ([Fli28], p. 636).

5.2.2 Relationalism and operationalism

Flint and Ruark were explicit about their operationalist motivation: they both regarded measurable quantities as ontologically primitive, preceding any ontological claim about the structure of abstract space. This view enjoyed wide support in the 1920s, for example, from Eddington:

> I should be puzzled to say off-hand what is the series of operations and calculations involved in measuring a length of 10^{-15} cm; nevertheless I shall refer to such a length when necessary as though it were a quantity of which the definition is obvious . . . I may be laying myself open to the charge that I am doing the very thing I criticize in the older physics – using terms that have no definite observational meaning, and mingling with my physical quantities things which are not the results of any conceivable experimental operation . . .
>
> *([Edd23], pp. 6–7)*

The fundamental length was seen as limiting any possible theoretical description of physics of shorter length scales. Eddington saw it as a cure to the puzzles of quantum phenomena:

> Indeed it has been suspected that the perplexities of quantum phenomena may arise from the tacit assumption that the notions of length and duration, acquired primarily from experiences in which the average effects of large numbers of quanta are involved, are applicable in the study of individual quanta.
>
> *([Edd23], p. 7)*

Ruark and Flint drew more metaphysical lessons on the status of space and time (and thereby of spatiotemporal relations) at such a scale:

> . . . This result [the existence of a lower bound on the possibility of resolving wavelengths with scattering experiments – AH] indicates a definite limitation which affects any attempt to study the structure of an individual electron by the use of light quanta. [. . . it] do[es] not cripple the program of applying causal law and continuous analysis to sub-electronic problems; but in so far as the assumptions used are valid in this domain they deprive such a program of experimental support – the sine qua non of physics – and throw the problem into the realm of speculation.
>
> *([Rua28], pp. 326–327)*

The principle of minimum proper time is itself of the nature of an exclusion principle, limiting the region of application of the quantum equation. The principle suggests that this equation must not be applied when the region is one where changes occur in a distance of h/m_0c; in fact, it suggests that to speak of changes of this kind in association with particles does not correspond to physical reality, when applied to cases of very rapid motion. This being the case our development of the equation shows it to be part of a theory which sounds a note of warning when we approach the confines of its domain.

([Fli36], pp. 443–444)

This operationalist account was also explicit in Arthur March's theory of fundamental length (see also Section 4.3.2). March proposed [Mar36] replacing the ordinary metric $ds^2 = g_{\mu\nu}dx^\mu x^\nu$ with $ds = \sqrt{g_{\mu\nu}dx^\mu x^\nu} - \rho$ where ρ is the fundamental length. The distance between two world points P and Q would then be given by

$$s = \int_P^Q \sqrt{g_{\mu\nu}dx^\mu x^\nu} - \rho \tag{5.1}$$

and is valid only in the macroscopic domain, whereas for P and Q very close together, ds would become negative, and therefore not interpretable as a distance.

March's theory was an operationalist theory of length measurements. As far as the structure of spacetime was concerned, March was unclear about it. Presumably he held it to be a continuum, or at least he saw his theory about physical (observable) space as consistent with the mathematical continuum. He refrained, for example, from suggesting replacing the differential calculus with difference equations, as Heisenberg and Ruark did (see also [Kra95]).

The idea that physics should only concern itself with measurable quantities was advocated by Heisenberg [Hei43], and more recently by G. Chew in his *S*-Matrix theory and the bootstrap principle [Che63]. Similar to March, Chew displayed complete agnosticism with respect to the continuum, although he argued for the use of only finite (measurable) quantities in microphysics (in the *S*-Matrix case these quantities were the momenta of the particle).

Stripped of any other philosophical view of spacetime, an operationalist approach such as that advocated above may be attacked on the grounds that it spills the baby with the water, as according to its own predictions, the structure of spacetime becomes a purely metaphysical question. In a printed comment to one of his papers, for example, Flint was criticized by one angry Mr. J. Guild:

The view that space and time lose their significance below certain limiting values is a purely metaphysical escape from the difficulties which confront mathematical analyses, of the types so far developed, when carried beyond these limits. Unless there is some conceivable experiment which would distinguish discontinuities in space or time, such discontinuities cannot exist even as a pragmatic hypothesis. . . . We must avoid the tendency, when confronted with analytical difficulties, to invest nature with properties whose only function is to conceal our troubles. Eddington has warned us, in connexion with cosmic problems, of the danger of constructing a wall at the edge of the universe over which we can throw our unsolved problems. The quantum physicist does not carry his difficulties to the edge of the universe to dump them over the wall, but wants apparently to stuff them into holes in time and space. It is better to leave them lying about as a reminder that there is still work to be done in tidying up the unsolved problems of analysis.

([Fli38], pp. 94–95)

The accusation is unfair on several counts. First, as we have argued in Chapter 3, the operationalist view does not necessarily entail that the question of spatial discreteness is purely metaphysical, or transcendental. Second, Flint was careful to formulate his view as an empirical hypothesis with concrete predictions ([KC94], pp. 450–454). He argued, for instance, that it explained the number of elements in the periodic table; that it explained the proton–electron mass ratio (predicting along the way the mass of the hypothetical neutron); and that it set an upper limit for the energy of cosmic radiation. The choice of the first two empirical "consequences" may have been unfortunate. Although numerological speculations were quite popular at that time, most physicists realized that it was just too easy to obtain spectacular confirmation between empirical data and sophisticated combinations of the constants of nature. The upper bound on energy, on the other hand, was testable *in principle*.

Finally, and in contrast to March, rather than being agnostic about the structure of spacetime, Ruark and Flint advocated, albeit implicitly, a relationalist view of spacetime (for a modern exposition of relationalism and a detailed analysis see [Bar82] and Chapter 6 in [Fri83], respectively). In a nutshell, relationalism amounts to the denial of "empty spacetime regions": any claim about the spatiotemporal structure of a given spacetime region is to be construed as a claim about the spatiotemporal properties and relations of the concrete physical objects and events that are "in" that region. On this view, spacetime is not an independent ontological entity, a "container" for physical objects, but a convenient way of saying something about the spatiotemporal relations of concrete physical objects. This relationalism defuses attacks such as the above: there are no holes in time and space into which the operationalist stuffs her difficulties; there are only material bodies.

Spacetime relationalism is also implicit in Schrödinger's views on the unification of quantum theory and the special theory of relativity [Kra92b]. In Schrödinger's case, this relationalism was motivated by his lack of confidence in the precedence of the specific geometrical structure of mathematical space (which he regarded as a mere idealization) over the structure of physical space, described by the material bodies whose dynamics is governed by quantum theory. That Schrödinger was not swayed by the formal requirements of mathematical space and its symmetries, for example, the continuous Lorentz symmetry, is evident from his alternative to Dirac's hole theory ([Kra92b], pp. 329–330). His reasoning was that these geometrical symmetries were couched in a classical worldview, and their applicability to quantum theory should not be taken for granted. Concepts such as rigid rods or clocks presuppose the classical ability to localize them with infinite precision, which is lost when one moves to quantum theory:

Classical mechanics allows one to think of any forces between point masses even if such point masses and such forces do not really exist. Using an appropriate potential one could bring two point masses as close as one likes. In quantum theory this is not allowed. From one energy level to the next there is always a finite gap . . . It seems that quantum theory resists the idealization that one would have to allow in order to make physical localization meaningful as in the mathematical case.

([Sch34], p. 519)

According to Schrödinger, quantum theory has exposed the difference, hidden in the classical case, between physical space and mathematical space. The latter is a mere idealization; the former is the space in which natural processes take place, and since these (quantum) processes do not allow infinite precision, the fact that ideal (continuous) geometry is inapplicable to this space should not come as a surprise.

The idea here is that reference frames (the rods and clocks) which are used to fix the order of physical events should consist of real physical objects, and so the position of a particle, as given by its spatiotemporal coordinates, should be identified with the particle's position operator. Since quantum mechanics imposes limitations of measurement on such an operator, this limitation is at the same time a limitation on the structure of physical space.

Consistent with these ideas, Ruark [Rua31] proposed a calculus of difference equations to replace the standard, differential calculus, arguing that the latter is an unjustified extrapolation from the finite set of observations of finite (at most rational) magnitudes we can ever make. As we have seen (Section 4.3.2), Heisenberg, too, had proposed such a modification. It would take another generation for these ideas to be developed further by mathematical physicists in the 1950s and 1960s.

The relationalism behind the notion of absolute uncertainty is evident also in what, to my mind, is the first attempt to tackle head on the tension between fundamental length and Lorentz contraction. This attempt is credited to Wataghin [Wat30b], who also proposed an alternative derivation to Flint's and Ruark's limits on resolution [Wat30a]. It exemplifies Wataghin's extremely concise – sometimes even cryptic – style of prose, and may even be regarded as a symptom of a "rush to print."[2] Here it is in English:

But this difficulty [the conflict with Lorentz contraction – AH] disappears (as we will show in a forthcoming paper) if one considers the following circumstances: in deriving (1) [the minimal uncertainty in length measurement $\lambda = h/mc$ – AH], one must, as is well known, make a precise separation of the object to be observed from the means of

---

[2] In [Wat75], p. 31, Wataghin confesses that at that time issues of priority were high on everyone's agenda.

observing. Besides, it is necessary to think of every system of reference as being realized by the measurement apparatus, and so to count the system of reference itself among the experimental devices. Therefore one must revise the concept of the Lorentz transformation from the quantum-theoretical point of view.

We come to the result that one surely has to differentiate between two kinds of Lorentz transformations: the Lorentz transformation in the usual sense between the variables x_i and x'_k, which are considered c-numbers, and the "physical" Lorentz transformation, which is brought about by a change in the means of observation. In the first case, it is a matter of a change in the name with which we designate different points in space-time. In the second case, which is the only one of significance in quantum theory,

$$(2) \quad x_i = \sum_k \alpha_{ik} x'_k,$$

the x_i, x'_k and α_{ik} are to be considered as "observable" or quantum values (matrices). A closer analysis shows that the relationship (1), and likewise the Heisenberg–Bohr uncertainty relations, are invariant with respect to such Lorentz transformations. It turns out, by the way, that the limit of precision (1) holds for any given velocities of the particles, even when the relativistic changeability of mass is taken into consideration.

([Wat30b], pp. 650–651)

To decipher Wataghin's cryptic way out of this conflict, one must recall the operationalist and relationalist motivations behind the notion of absolute uncertainty. In this framework, the spatiotemporal relations, and in particular the position of a particle (x), fixed by a coordinate system, supervene on the material bodies with which one fixes the reference frame and with which one makes spatial and temporal measurements. Wataghin is saying here that while the former x may be described by an abstract c-number, the latter $\hat{x}$ is a dynamical object, governed by quantum theory, and hence is not a number but an operator, described by a matrix. A (formal, but unphysical) Lorentz transformation of the former is just a passive transformation "in label." A physical Lorentz transformation of the latter operator is still possible, inasmuch as Heisenberg's uncertainty relations are Lorentz invariant.

If one takes Wataghin literally, he is simply noting that the new spatial resolution limit he has derived is consistent with Lorentz invariance, in the sense that a boosted observer will also be subject to it in her own rest frame (which forms the measurement apparatus for the scattering experiment she performs). This means that the fundamental length is an observer independent magnitude, which then allows Wataghin to generalize Flint's and Ruark's results to systems in motion.

This interpretation is quite benign, but it does tell us that consistency between fundamental length and Lorentz invariance can be achieved if we take the former to represent a limitation on *proper length* resolution (in one's own rest frame), and not as an actual physical rod of a certain minimal size viewed by two or more observers. As we shall see below, the standard reply today in the quantum gravity community (suggested by Snyder [Sny47a] and recently by Rovelli and Speziale

[RS03]) to worries about inconsistency between the Lorentz contraction and the notion of fundamental length, relies on a similar intuition.

5.2.3 Born's reciprocity

Considerable attention has been given in recent years to the notion of "duality" [Wit97, CON10]. One of the first to propose this notion was Max Born [Bor38a].[3] In an attempt at unifying quantum theory with the principle of general covariance that underlies the general theory of relativity, Born observed that the motion of a free particle is completely symmetric in the two four-vectors x and p.[4] The transformation theory of quantum mechanics extends this "reciprocity" systematically:[5] in a representation of the operators x^k, p_k, in the Hilbert space for which the x^k are diagonal, the p_k are given by $\frac{\hbar}{i}\frac{\partial}{\partial x^k}$ and vice versa. Any equation in the x-space can be transformed into another equation in the p-space:

$$\phi(p) = \int \psi(x) \exp\left[\frac{i}{\hbar} p_k x^k\right] dx. \tag{5.2}$$

This reciprocity is extendable to cases where external forces are operating on the particle.

Born then observes that the application of reciprocity to general relativity, where the line segment

$$ds^2 = g_{kl} dx^k dx^l, \tag{5.3}$$

describes phenomena on cosmic scales (g_{kl} is the metric tensor), leads to a reciprocal line segment in p-space

$$d\sigma^2 = \gamma_{kl} dp_k dp_l, \tag{5.4}$$

which can be used to define a metric on that space, one which is not directly connected with the metric tensor g_{kl} in the x-space.

This assumption of independence, of course, breaks down in classical mechanics, where $p = m\dot{x}^k$ (m is the rest mass and the dot is time differentiation), but

---

[3] It should be remarked, for the sake of historical precision, that Wataghin derived a dual metric on momentum space already in 1937 [Wat37b]. That for him momentum space was curved and possessed some non-trivial geometry is evident from (a) his ideas about the high momentum cutoff and (b) the fact that he referred to it as such in his later publications (e.g., [Wat72], p. 610). However, Wataghin never developed this idea into a coherent theory, and the extent to which this dual space played a conceptual role in his attempt to unify quantum theory and general relativity remains unknown.

[4] Born was no stranger to ideas about minimal length and its role in the attempts to remove the singularities of field theories. As early as 1934 he suggested a non-linear field theory with a high momentum cutoff (see [BI34] and Section 4.2.2). For him the minimal length (at the scale of the electron's radius) was closely connected with the fine-structure constant and the magical number 137 [Bor35].

[5] Born traces the word "reciprocity" to the lattice theory of crystals where the motion of the particle is described in the p-space with help of the "reciprocal lattice."

Born emphasizes that classical mechanics is only one limiting case, i.e., when we consider particles of low energy and momentum (as compared with $h\nu_0$ and h/λ_0, where $\lambda_0 = c/\nu_0$ is the Compton wavelength). Another limiting case is the domain of high energy physics, which has to do with very small regions of space and time (compared with λ_0 and $1/\nu_0$), but with unrestricted amounts of energy and momentum. Born then argues that reciprocity demands that we should not give precedence to ds^2 in all cases; it is applicable only in the appropriate low energy limiting case, as much as $d\sigma^2$ is applicable only for high energy phenomena (Born calls the former "the molar world" and the latter "the nuclear world"). Most importantly, in cases which involve both scales simultaneously with an equal measure, the concept of metric becomes inapplicable, analogously to the uncertainty principle.

The consequences of reciprocity are far reaching. First, in the same manner as g_{kl} satisfies differential equations which connect the curvature of x-space with the stress–energy tensor, one could now write equations for γ_{kl}, and treat p-space as curved. Next, as much as Einstein's equation of state admits a static solution corresponding to a closed (hyperspherical) world filled with matter of uniform density, where there exists an upper limit for the distance between two points, given by the radius a of the universe, the reciprocal equation of state in p-space admits a corresponding solution, and therefore, for certain systems, signifies an upper limit for momenta, determined by the radius b of the hypersphere in p-space (here Born mentions the cutoffs of Wataghin [Wat34a] and March [Mar36], but remarks that their cutoffs were not motivated by relativistic considerations such as his own was).

Born then notes an important by-product of his suggestion, namely the elimination of divergences in QED by the upper bound on momentum (as the number of quantum states in a closed volume in p-space becomes finite – note the similarity with von Neumann's reasoning [vN36]). The question whether b is universal or only relative to the energetic system at hand (i.e., whether the fundamental length is minimal or just characteristic to the mass of the relevant elementary particle) is left open. Applying this to Fermi's version of QED, Born obtains numerical results for the radius of the electron and the corresponding value of b.

Born's reciprocity principle has empirical consequences. Applying it to the theory of heat radiation, one obtains a modification from Maxwell's equations for high frequency waves in a way that the relation of the pressure of light to the density of energy is consistent with an upper bound on momentum. Further deviations from the caloric properties of gases are predicted for very high temperatures, and in a later paper [Bor38b] more numerical predictions are derived for scattering phenomena and nuclear radii, and for the mass of the newly discovered meson.

Ten years later, in a contribution to a volume that celebrated Einstein's 70th birthday, Born returned to his principle of reciprocity [Bor49] and motivated it operationally. How, he asks, can one apply the invariance principle of the spacetime interval – a principle with empirical consequences, measured by macroscopic instruments, rods and clocks – to atomic scales? In other words, how can one determine the distance between two spacetime events if the instruments are macroscopic and the distance is of atomic scale? The lack of observability of atomic scale distances, and the accumulation of characteristic lengths with the ongoing discoveries of relativistic particles such as the meson both support, according to Born, his principle of reciprocity. He then laments that the idea of reciprocity found very little resonance among contemporary physicists. The few collaborators he had succeeded in harnessing for the development thereof have nevertheless made progress in formalizing general conservation rules and field quantization with an upper bound on momentum.

One consequence of this formalization is that while for macroscopic bodies ordinary special relativity holds, for high momentum particles this is not the case, and one must modify the spacetime metric to represent its dependence on the metric in p-space. Another consequence is the modification of the Coulomb law – suggested by Born and Infeld [BI34] for independent reasons of eliminating the divergence of the self-energy of the classical electron. In a short communication in *Nature* a year later [Bor50] Born cited Yukawa's non-local field theory as yet more evidence for the methodological fruitfulness of his principle of reciprocity.

Born's insight according to which the x-space and the p-space are subject to geometrical laws of the same structure, namely a Riemannian metric, was cited only a few times in the two decades after its inception, and was rediscovered only three generations later, in recent attempts to construct a theory of quantum gravity. In contrast to Wataghin, for whom momentum space had precedence over position space, as it allowed him to evade the tension between the high momentum cutoff and the notion of a dimensionless point, Born appears to have seen the duality as a symmetry relation, and gave no precedence to one space over the other. It is interesting that while the idea of reciprocity has reappeared recently in an approach to the problem of quantum gravity known as deformed special relativity [AC12a], proponents of that view have also given precedence to momentum space, whereas position space, and in particular the notion of locality ("an event in a given spatiotemporal point") are claimed to supervene on the properties of the former [ACFKGS11]. We shall discuss these issues in detail in Chapter 7.

5.3 Gravitons, measurability, and the Planck scale

5.3.1 Matvei Bronstein and Soviet theoretical physics in the 1930s

One of the first attempts to quantize the gravitational field belongs to the Russian theoretical physicist Matvei Bronstein (see [GF94], pp. 83–121 and references therein). Already in 1930, following Heisenberg and Pauli [HP29], Rosenfeld

[Ros30] analyzed a system comprising an electromagnetic field and a weak grav-itational field. The latter could be approximated by linearized Einstein equations, which meant that one could ignore its geometrical meaning and treat the system as two fields, a vector and a tensor, in a flat spacetime. But Bronstein was motivated by a more general picture, in which physics should be characterized by the three constants, Newton's G, Einstein's c, and Planck's $\hbar$. QED was a unification of c and $\hbar$, and Bronstein sought for a theory that would involve all three constants, that would delineate the quantum limits of the general theory of relativity.

Bronstein's treatment of the quantum theory of the weak gravitational field yielded a calculation of the energy radiation caused by the emission of a gravita-tional quantum. In the classical limit ($\hbar \to 0$) his theory approximated the classical theory. This methodology – the derivation of Newton's gravitational force from a quantum theory of gravity – was analogous to that adopted by Fock and Podolsky a few years earlier, when they derived the Coulomb force from QED. The result was thought to indicate that a quantum theory of gravity was not only possible, but also indispensable, and it was especially significant in Bronstein's contem-porary climate, where the gravitational field – with its geometrical meaning and non-linear character – was regarded as different than other matter fields. This reluctance to quantize the gravitational field was not unjustified: the approach that worked so well for the other forces of nature did not seem applicable to gravity (see Section 5.6).

The 1930s were also rich with controversies on the consistency of QED, espe-cially with respect to the measurability of field quantities (see Section 4.3.1). Bronstein was the first to extend the debate to the quantum gravitational domain.[6] Using, again, the weak field approximation, he identified the correct Hamiltonian for the gravitational field and proceeded to apply Heisenberg and Pauli's method, outlining the commutation relations ([GF94], pp. 102–105). He then followed Bohr and Rosenfeld [BR33], introducing a minimal volume of the test body that is used to measure the gravitational field intensity (in terms of the Christoffel symbols). His analysis showed that in order to make a maximally exact measurement of the gravitational field intensity in a given volume, we need to employ the test body of largest mass density possible; if concentrated too much beyond its Schwarzschild radius, this mass will collapse to what we call today a black hole.

This conclusion demonstrated, according to Bronstein, the profound difference between the electromagnetic and the gravitational fields: QED does not limit the charge density – the higher it is, the more accurate the field component measurement is, and one can get arbitrarily close to it – while in a future theory of quantum gravity

---

[6] Bronstein's extension of the Bohr and Rosenfeld argument on the measurability of field quantities to include gravity, as well as the arguments that iterated his move in the 1950s all rely on the aforementioned "disturbance" view of the uncertainty principle. We shall discuss this view in Section 5.6 below.

there is an *absolute limit* to the mass density, beyond which the general theory of relativity loses its applicability.

the gravitational radius of the test-body ($G\rho V/c^2$) used for the measurements should by no means be larger than its linear dimensions ($V^{1/3}$); from this one obtains an upper bound for its density ($\rho \leq c^2/GV^{2/3}$). Thus, the possibilities for measurements in this region are even more restricted than one concludes from the quantum-mechanical commutation relations. Without a profound change of the classical notions it therefore seems hardly possible to extend the quantum theory of gravitation.

([Bro36], translated in [Gor05])

To continue the analogy with the Bohr and Rosenfeld measurability analysis, it seems that in contrast to QED, where only two physical constants, c and $\hbar$, are required for the theoretical apparatus, and therefore this apparatus is independent of the ideas of the atomic structure of matter (as these two constants are insufficient for the construction of a universal length of an interval), in a theory of quantum gravity three universal constants, $c, \hbar$, and G, can make up the universal length of the theory.

In modern terms, Bronstein exposed the Planck scale as the limit of applicability of the general theory of relativity: if one tries to measure the gravitational field as precisely as possible in a volume as small as possible, then one can obtain, for example, the indeterminacy of the metric:

$$\Delta g = l_{Planck}/cT, \tag{5.5}$$

where $l_{Planck} = (hG/c^3)^{1/2} = 10^{-33}$ cm is the Planck length, and T is the measurement duration.

Bronstein understood that "the absolute limit is calculated roughly" (in the weak-field framework), but he believed that "an analogous result will be valid also in a more exact theory." In 1936 he formulated the fundamental conclusion as follows:

The elimination of the logical inconsistencies connected with this requires a radical reconstruction of the theory, and in particular, the rejection of a Riemannian geometry dealing, as we have seen here, with values unobservable in principle, and perhaps also the rejection of our ordinary concepts of space and time, replacing them by some much deeper and nonevident concepts. Wer's nicht glaubt, bezahlt einen Taler.

([Bro36], translated in [Gor92])

The same German phrase concludes one of the Grimm brothers' stories, "The Brave Little Tailor."

Matvei Bronstein, one of the trailblazers of quantum gravity, ended his career in a Leningrad prison in February 1938, where he was executed at the age of 31.

5.3.2 *Wataghin and Schönberg: a forestalled attempt*

Gleb Wataghin's work on fundamental length, as we have seen, was downplayed by his contemporaries, especially by Heisenberg. This negative attitude – partly explainable by Wataghin's minimalist style and condensed, sometimes cryptic, writing – is manifest in Wataghin's attempt to construct a quantum theory of gravity in 1936. By that time Wataghin was already in Brazil, at the University of Sao Paulo, where, together with his colleague Mario Schönberg, he apparently developed the seeds of a theory of quantization of the gravitational field. We learn of this attempt from a letter Schönberg sent on November 6, 1936, to one of his friends (source in Portuguese, translation mine):

.... I have started working on a quantum theory of gravitation ... together with prof. Wataghin we are constructing a theory that generalizes current quantum mechanics, introducing a lower bound Λ on observable spacetime intervals. On October 21 we sent a note to *Lincei* and a letter to the *Physical Review*. In this work we discuss the modification of the operators that represent spatio-temporal coordinates and the commutation rules of Heisenberg. In a second note we examine the Dirac–Schrödinger equations and the theory of quantum electrodynamics; the theory yields an exact value of the electron mass, setting Λ to be equal to the electron radius, eliminating, in a very simple way, one of the greatest difficulties of current quantum theories. In a third note we try to construct a theory of quantum gravity based on different principles than those of the theory of the electromagnetic field.

([Sch36], p xxiii.)

The paper was sent to the *Physical Review* but was never published. The ledgers of the journal at the American Institute of Physics have a record of the submission; it was refereed by H. P. Robertson (this seems to have been a reasonable editorial decision at that time, given Robertson's famous generalization of the uncertainty relations from 1929), but was rejected for an unknown reason and with no comments. This was the only time Wataghin and Schönberg attempted to publish a collaborative work, although they did sometimes give credit to each other in their papers.

Without additional archival evidence,[7] it is hard to reconstruct the ideas that Wataghin and Schönberg had in mind. Schönberg later published work focused on QED and the electron radius, and not on quantum gravity (although one paper of his discusses discrete geometry [Sch57]). Wataghin, on the other hand, published three short notes on the subject in *Lincei* and *La Ricerca Scientifica* – the bulletins of the Italian Academy of Science and Research Council, respectively – in September

---

[7] In his interview with Cylon Eudóxio Silva in 1975, held in Rio de Janeiro ([Wat75], p. 35), Wataghin expresses his view that general relativity may be a macroscopic theory, and that its quantization may lead to different physics at short distances, but says nothing more about the subject. His colleagues in Turin, for example, were unaware of his failed attempt at quantizing gravity, and it seems that he remained silent about it for many years.

1936, in May 1937, and in August 1937, where he essentially repeats the ideas of absolute uncertainty and fundamental length. Wataghin's short excursion into quantum gravity appears to consist solely of these three short notes, although as late as 1972 [Wat72, Wat73] he mentions some of the ingredients of this theory, for example, absolute uncertainty, fundamental length, modified commutation relations in momentum space, and thermodynamic reasoning.[8]

In the first two notes [Wat36, Wat37b], Wataghin draws an analogy between the quantization of the electromagnetic and gravitational fields, and mentions neutrinos as possible carriers of the gravitational force. The cryptic note ends with a conjecture that spatiotemporal operators should be subject to uncertainty relations because of fluctuations inside a minimal spatiotemporal "volume" – fluctuations that Wataghin connects with *Zitterbewegung* ("trembling motion"),[9] the rapid motion of elementary particles, in this case, of the neutrinos. In the third note [Wat37a], Wataghin iterates his intuition – now based on formal considerations regarding Dirac's equation in a curved spacetime – that a gravitational particle of spin one-half must be introduced in order to maintain conservation of momentum (in the same way that the neutrino maintains conservation of energy in Fermi's theory of β decay).[10] He then derives the metric for the unified theory, and a dual metric for momentum space, but tells us nothing in print about the conceptual importance of the latter.

Wataghin was not alone in his ideas. Following the example set by QED and the photon, and by Fermi's theory of β decay and the neutrino, a few physicists sought a mediating particle for the gravitational force. The name "graviton" was coined in 1934 by two Russian theoretical physicists, Blokhintsev and Gal'perin (for a translation see [GF94], pp. 96–97). The two speculated in print about the possible identification of the graviton with the neutrino. A connection between the neutrino and the quantum of gravitational waves was also entertained by Bohr in 1933 and reported by Gamow as "an exciting theoretical possibility" ([Gam62], p. 143). Even as late as 1962, Feynman mentions this idea in his famous lectures on gravitation [FMW03], and, for the sake of motivating the notion of a quantum of gravity, discusses whether the exchange of one neutrino could serve as an example for a force anything like gravitation.

---

[8] Three later papers of Wataghin [Wat38, Wat43, Wat44], the first published in *Nature* and the other two in *Physical Review*, concern quantum theory and *special* relativity. In these papers he only discusses his earlier suggestion of a high momentum cutoff and some ways around the non-local effects it yields.

[9] The idea of *Zitterbewegung* was first proposed by Schrödinger in 1930 as an alternative to Dirac's hole theory, see [Kra92b], pp. 323–329 and references therein.

[10] With the clarity of hindsight, Wataghin's reasoning seems, today, unwarranted: quantum field theory, if extended to include gravity and the graviton as a mediating particle for the gravitational force, requires it to be a massless spin two particle (see, e.g., [FMW03], pp. xxxi–xxxvi). One should recall, however, that Wataghin was working before the discovery of the electro-weak theory and later gauge theories.

5.4 Non-commutative geometry

The reconstruction of Wataghin's speculative "theory" of quantum gravity, fragmented as it is, demonstrates the direct conceptual link that connects the relational-operational motivations behind absolute uncertainty, via the limitation on measurability of field quantities – in QED and in a possible quantum theory of gravity – with the idea of non-commutative geometry. While many of Wataghin's ideas are being rediscovered today, we know of no direct link that can be traced back to him from current research.[11] Heisenberg, on the other hand, is much more credited in the quantum gravity community of the twenty-first century. But even in his case, it took almost a generation for the link to be established in print [Wes02]. Like pollen in the wind, the memes were transferred: from Heisenberg in a letter to Peierls written in 1930 ([Pau05], Vol. 2, p. 15), on to Pauli and via him to his great admirer Oppenheimer, and finally to Oppenheimer's student, Hartland Snyder [Sny47a]. And so, almost 17 years later, non-commutative geometry was put on mathematically rigorous ground, albeit without mentioning Heisenberg's original speculation ([Pau05], Vol. 3, p. 348), namely, that a fundamental length can arise from letting two position operators – and not only position and its conjugate momentum – be non-commuting, i.e.,

$$[\hat{x}^{\mu}, \hat{x}^{\nu}] \neq 0. \tag{5.6}$$

5.4.1 Hartland Snyder's quantized spacetime

The American physicist Hartland Snyder is famous mainly for discovering, along with Oppenheimer, what today we call "black holes," by calculating the gravitational collapse of a pressure-free homogeneous fluid sphere [OS39], and for laying down, along with Ernest Courant and M. Stanley Livingston, his colleagues from Brookhaven laboratory, the foundations of accelerator physics [CLS52]. In between these two achievements, in 1947, Snyder presented one of the first mathematically coherent and self-contained papers about spatiotemporal discreteness.

Snyder begins his paper [Sny47a] by registering his dislike of the way spatial discreteness has been introduced into physics in the past, namely, via high momentum cutoffs that seem to be impossible to reconcile with relativistic invariance. His

---

[11] One possible reason has to do with Wataghin's style. In contrast to his contemporaries, e.g., Flint, Ruark, March, Heisenberg, and Born, who expressed their views at length and were quite prolix in their publications, Wataghin restricted himself to short notes with few equations, and resorted to an extremely economical, almost cryptic, prose. Another circumstantial reason has to do with Wataghin's "choice of venue" for these topics that may have been detrimental to his influence, publishing as he was almost exclusively in *Il Nuovo Cimento* and *Lincei*, journals whose international impact had been steadily decreasing throughout the years. On the other hand, with the exception of two papers [Wat43, Wat44] which mention his high momentum cutoff, all of Wataghin's published work in the *Physical Review* concerns the far less speculative physics of cosmic rays.

abstract, on the other hand, promises a Lorentz invariant discrete spacetime. Note here that many who introduced the said cutoff, for example, Wataghin, March, and Feynman, to name a few, did so relativistically, to the extent that they preserved Lorentz invariance. Violation of the latter, as we have seen, characterizes a lattice structure, but the aforementioned cutoffs were not directly associated with this structure, at least not in all cases. Presumably, then, what Snyder had in mind that could not be reconciled with relativistic invariance was the breakdown of causal order that this cutoff entailed.

Faithful to his abstract, Snyder then notes, and shortly after also proves ([Sny47a], p. 40), that Lorentz invariance of the spatiotemporal coordinates does not require that these coordinates take on a continuum of values simultaneously. If one drops this requirement, one can represent the coordinates of a particular Lorentz frame as Hermitian operators. What remain invariant under Lorentz transformations are now the *spectra* of these operators,[12] and while the standard (continuum) case satisfies this requirement, it ceases to be the sole option: a discrete spacetime with a fundamental unit of length can satisfy it too.

The price one pays for introducing the latter, however, is that the spatiotemporal coordinates no longer commute, as they do in the continuum case.

The fundamental length a enters into the formalism in the definition of the spatiotemporal operators, whose eigenvalues are integral multiples of a. When $a \to 0$ the discrete manifold reduces back to its ordinary, continuous, form. This relation also holds for the space- and time-displacement operators (momentum and energy) Snyder defines, whose commutation relations with the spatiotemporal coordinates are modified, for example,

$$[x, p_x] = i\hbar[1 + (a/\hbar)^2 px^2], \tag{5.7}$$

and

$$[t, p_t] = i\hbar[1 - (a/\hbar c)^2 pt^2] . \tag{5.8}$$

If all the components of the momentum are small compared to $\hbar/a$ and the energy is small compared to $\hbar c/a$ (or if $a \to 0$), then these commutators approach those which are given by ordinary quantum mechanics. In other words, that these new commutators differ only for large values of the momentum (when $|p| > \hbar/a$) implies, according to Snyder, that a field theory based on a quantized spacetime will essentially yield the same predictions as an ordinary, continuum based, field theory, for all processes that involve small momenta, but will disagree

---

[12] The spectrum of a spatiotemporal operator is composed of the possible measurement results of the corresponding quantity; the operators x, y, z, t shall be such that the spectra of their Lorentz transformed operators x', y', z', t' formed by taking linear combinations of x, y, z, t, which leave the quadratic form of the spacetime interval invariant, shall be the same as the spectra of x, y, z, t, see [Sny47a], p. 39.

with the latter for processes that involve large components of momenta. Presumably, processes that involve self-energy and vacuum polarization are of the latter kind.

Snyder proves invariance of these non-commuting coordinates (in the sense above of preserving their spectra) only with respect to rotations that leave the origin fixed and – similarly to the coexistence of discrete spectra of angular momentum with continuous rotational transformations – shows that spatial discreteness need not violate continuous Lorentz rotation. He admits, however, that, as expected, a continuum of translations is not admissible in this space – if it were, the manifold would have been a continuum.[13] Translations of the origin may be introduced, but it should not be expected, nor is it possible in a quantized spacetime, to find sharply defined translational values for all four coordinates simultaneously. When setting up the relation between two Lorentz frames in this space one cannot bypass the limitation on precision which is imposed by the commutation relations between the coordinates.

Snyder returned to his quantized spacetime in the same year [Sny47b], to investigate whether relativistically invariant field equations could be introduced into a discrete background in such a way that they would be solvable. The question is not trivial. In continuous spacetime, the field equations govern quantities which are functions of the spacetime coordinates. Since the latter are continuous, the former can be written as partial differential equations. In quantized spacetime the spatial coordinates do not commute, hence the difficulty in defining functions for non-commuting variables. Moreover, since time is discrete, partial derivatives are not definable in the ordinary sense.

Snyder follows two intuitions. First, he regards field quantities not as functions of spacetime coordinates, but as operators on the Hilbert space on which the spacetime operators operate. Next, he uses his definition of displacement operators in [Sny47a] and shows that if $A(x, y, z, t)$ is a field quantity of continuous spacetime, and if the term $\partial A/\partial x$ appears in the field equation, it can be replaced by $i[p_z, A]$ in the transition to quantized spacetime ([Sny47b], p. 68).

These intuitions yield an application to the vacuum Maxwell equations that is both consistent and relativistically invariant. The problem, thus, is not formal, it is a problem of meaning: what is the physical meaning of these commutator equations, and can operators satisfying them be found? Snyder answers only the second question, by constructing such operators, which are expressed in terms of the wave number-frequency components of the coordinate operators, analogously to the usual Fourier analysis of fields. He notes that the essential change in the

---

[13] In this context see an interesting suggestion [Yan47] to include translations in a quantized spacetime by allowing it to be curved (e.g., a de Sitter space which is a pseudosphere in five dimensions).

transition from continuous to quantized spacetime is a change in the definition of the scalar products of the vectors in the Hilbert space ([Sny47b], p. 71).

5.4.2 Follow-ups

Within the quantum gravity community, Snyder's papers are today considered seminal, and Snyder himself is often cited (currently with over 600 citations, most of which are from the 1990s onwards) as being the first to introduce Lorentz invariant spatial discreteness (see, e.g., [Hos13], p. 8). From a historical perspective, as we have seen in the last two chapters, similar relational and operational ideas, for example, uncertainty in position measurements and the consistency thereof with Lorentz invariance, were already expressed by Wataghin and March, who are far less credited in the quantum gravity community. One reason for this may be that, in contrast to Wataghin and March, who remained agnostic about the structure of spacetime, Snyder was explicit about the ontological interpretation of the fundamental length as signifying a discrete feature of spacetime itself, and not as merely representing an epistemic limitation on spatial resolution.

Shortly after their publication, several physicists used Snyder's papers as an anchor for expressing their own ideas about fundamental length. Here we shall mention a few of them.

Nathan Rosen's statistical geometry

Nathan Rosen, the "R" from the famous EPR paper (1935), starts [Ros47] from an operationalist stance, similar to that of March [Mar36], viewing the fundamental length (which he designates as the radius of the electron) as a limitation on measurement resolution. To describe this limitation, Rosen proposes a twofold mathematical description. The first, ordinary continuous spacetime, he calls the "abstract space"; the second, the one that describes our actual observations, he calls "observable space." The justification for that second space is strictly operational: the fundamental length delimits the ability to measure a spatial coordinate with infinite precision, or, as Rosen puts it, "there are no infinitesimally small measuring rods in nature," and, similar to March, for each point in abstract space there is a corresponding "volume," whose radius is a, which represents the measurement error.

A similar uncertainty, recall, was suggested by March, and appeared in the same formal way as a normal Gaussian distribution in Eddington's *Fundamental Theory* [Edd46] – Rosen was apparently unaware of this, and in a note added in proof acknowledges it was brought to his attention after publication. We have also encountered it in Ross' [Ros84] proposal for an alternative structure of the line segment (Section 2.5). Rosen developed the idea further to a geometry that deals with small regions instead of points. These regions are elementary volumes associated with mean values of the spatial coordinates, whose physical role in the "observable space" is equivalent to a spatial coordinate in "abstract space." While the value of a physical quantity at a point in "abstract space" is unobservable, only a physical quantity associated with such a "point" (volume) in the "observable space" can be observed.

In formulating dynamical laws on "observable space," Rosen suggests keeping them as close as possible to their original form on "abstract space," and shows this can be done by using an averaging process. As far as the Lorentz invariance of these laws is concerned, Rosen admits that the transformations from one reference frame to another in "observable space" may change as a result of introducing the fundamental length, but if one assumes Lorentz invariance on "abstract space," and if one knows how to transform from that space to the "observable space," one could discover these modified transformations.

The fact that one introduces an uncertainty only to spatial, and not to temporal, measurements is justified, according to Rosen, by the lack of operational meaning for temporal uncertainty. The spatial uncertainty represents the measurement error of the three coordinates of a point *at a given moment in time*; time serves as an external parameter that gives meaning to the measurement and identifies what is to be measured. If one introduces an uncertainty to a four-dimensional coordinate of spacetime, time can no longer serve as an external parameter, and, consequently it is not clear what one measures and how such a measurement can be repeated.

Rosen applies his statistical geometry of "observable space" to classical electrodynamics, and shows how the difficulties associated with the singularities are now removed. He suggests generalizing this view to quantum electrodynamics, and other quantum fields, by viewing the elementary particles not as points in "abstract space" but as elementary volumes in "observable space." Fifteen years later Rosen returned to this idea [Ros62]. Referring to Born's reciprocity (Section 5.2.3), he suggested that the symmetry between position space and momentum space may go deeper and arise from his statistical geometry. The consequences of applying these ideas to quantum field theory are non-locality and hyperplane dependence: field quantities are now defined on a simultaneity hyperplane, and not at a point.

Heisenberg on locality versus causality

Heisenberg's ideas about fundamental length and his attempts to incorporate them into particle physics were discussed earlier (Section 4.3.2). In this context, a paper he published in 1951 [Hei51a] is worth mentioning. There Heisenberg compares possible mathematical frameworks that can accommodate both his ideas about fundamental length, and the empirical facts that were accumulating about the creation of new particles in collision processes between elementary particles.

Heisenberg suggests describing field quantities as spinors $\psi(x)$, where x is the four-dimensional coordinate. The state of the system $\Psi(\sigma)$ is defined as a vector on Hilbert space, where σ is a three-dimensional surface on x (a simultaneity hyperplane), and its physical meaning is the standard quantum mechanical one, namely, the probability that a certain observable quantity has certain values on the surface σ. Ordinary QFT can be maintained for distances larger than the fundamental length (Heisenberg, recall, takes it to be of the order of 10^{-13} cm). The price for eliminating the divergences within it, as we have seen, is the sacrifice of relativistic causal order, and here Heisenberg repeats his observation [Hei51b] that such violations, while not necessarily confined to small scales, are nevertheless characteristic only to processes which involve high energy transfer, and these will be difficult to detect given that very little is known about them.

Snyder's quantized spacetime is next mentioned, alongside with Yukawa's non-local field theories, as alternative mathematical frameworks. Heisenberg's reservation regarding the two alternatives is telling:

In both cases it has hitherto not been possible to formulate the interaction of particles in a mathematically consistent scheme. But quite aside from these difficulties which may be overcome some day, one has the impression that such formalisms possibly abandon too much of "localization" as compared to the experimental situation.

([Hei51a], p. 21)

What seems to bother Heisenberg is that in both alternatives interactions do not occur *at a point*, while in collision processes, even in those which involve a very high energy, local interaction does seem possible, at least in one dimension, if one is willing to entertain the applicability of Lorentz contraction to distances smaller than the fundamental length. In three dimensions this localization may even justify the observation of particle creation – a wave packet smaller than 10^{-13} cm in all directions would by itself be a mixture of many different particles.

Heisenberg concludes that

it is scarcely the concept of localization that should be limited in order to find a consistent mathematical frame, but rather the concept of causality in the sense of action from a point only to the immediate neighborhood. It may be possible, however, to introduce deviations from causality in this sense only within small regions of the order 10^{-13} cm. The present scheme [i.e., his own suggestion – AH] introduces such deviations even in larger regions. This fact does not lead to logical inconsistencies; but it means that the description of the space-time changes of energy distribution would be limited by sharper restrictions than only by the usual complementarity between energy and time.

([Hei51a], pp. 21–22)

Flint and others

Other supporters of the notion of fundamental length reacted to Snyder's papers: Flint and Williamson [FW53], for example, restated Flint's ideas from a generation earlier [Fli28] and suggested an alternative to Snyder's coordinate operators. The mathematician Schild [Sch48a, Sch49a] suggested another mathematical consistency proof for the existence of a discrete spacetime that respects a subgroup of Lorentz transformations. His suggestion required only a discrete (integral) group of Lorentz transformations under which the lattice remains invariant, and led, in the case of a cubic lattice, to a smallest possible velocity parameter $v/c = \frac{1}{2}\sqrt{3}$, which is impossibly large to make the model of use for physical purposes. An improvement on this suggestion was offered by Hill [Hil55] who extended Schild's lattice from the integers to the rationals. Motivated by Snyder, Finkelstein [Fin49] developed a non-linear field theory that could incorporate particles as energy-density, momentum-density, and charge-density, interpreting the fundamental length not as a feature of spacetime but rather as the Compton wavelength of the particles.

It took almost four decades for the idea of quantized spacetime to make an impact on mainstream physics. Starting from the 1990s, Snyder's papers have become widely known, and the mathematics of non-commutative geometry first motivated, and then became widely applied in, theories of quantum gravity (see, e.g., [DN01] and references therein).

5.5 (Re)enters gravity

The idea that gravity introduces another limit on spatial resolution was first entertained by Bronstein in the mid-1930s. Almost a generation later, physicists began taking it seriously.

5.5.1 The Schwarzschild radius and the uncertainty of the metric

Among the first to iterate Bronstein's intuition was Oscar Klein [Kle56], who discussed possible connections between his unified theory of electromagnetism and gravity (Kaluza–Klein theory) and the quantization of the gravitational field. Early in his discussion Klein refers to a published paper he co-authored with Kosmos in 1954, and we find the following remark:

Let us assume that we have to do with a particle described approximately as a quantum belonging to a linear wave equation. Then by superposition we may make a wave package representing the particle confined to a volume of linear dimensions λ. If λ is small compared to the Compton wavelength of the particle the wave package will represent an energy $\sim \hbar c/\lambda$ and thus a mass $\sim \hbar c/\lambda$. Thus the difference in gravitational potential between the centre and the edge of the wave package will be $\sim G\hbar c/\lambda^2$ [where G is Newton's gravitational constant – AH] and will mean a negligible change of the metrics only if $\sim G\hbar c/\lambda^2 \ll c^2$, i.e., if $\lambda \gg \sqrt{G\hbar/c^3} \sim l_0$ [where l_0 is the minimal length – AH]. From this consideration it would seem to follow that the linear wave equation for the particle in question would break down when the wave length approaches the length l_0. The condition in question can also be expressed by stating that for λ approaching the length l_0 the gravitational self energy of the particle approaches the kinetic energy corresponding to its volume.

([Kle56], p. 61)

Shortly after, Klein attended a conference in Turin, convened by Wataghin and dedicated to non-local field theories. His contribution focused, again, on the gravitational origin of the minimal length [Kle57], where he repeats the argument that apart from the numerical combination of the three constants ($\hbar$, c, and G), the physical justification for this length l_0 comes from the limitation on the amount of mass which can be concentrated within a sphere of a given diameter d:

$$M < \frac{c^2}{4G}d. \tag{5.9}$$

When this bound is applied to a wave packet of diameter d corresponding to a quantum particle, and when d is smaller than the Compton wavelength of the particle, the energy (and therefore the mass) will be of the order of magnitude $\hbar c/d$, and the above inequality yields

$$d > \frac{l_0}{\sqrt{2}}. \tag{5.10}$$

And so l_0 is the natural limit on the region within which a particle can be confined.

Although not a proof, these considerations seemed to support the idea that the inclusion of gravity in QED would eliminate the divergences that plagued field theories. Other physicists were clearly convinced. Dirac, for example, who was known for his dislike of the renormalization program, found relief in this argument:

Now in considering a particle interacting with the gravitational field, the first thing one would think of would be a point particle, but here one runs into a difficulty, because if one keeps to physically acceptable ideas, one cannot have a particle smaller than the Schwarzschild radius, which provides a sort of natural boundary to space [...] because to send a signal inside and get it out again would take an infinite time. [...] Each particle must have a finite size no smaller than the Schwarzschild radius.

([Dir62], p. 354)

In the attempts to make this intuition rigorous, the role of quantum uncertainty was added to that of the gravitational limit. In a contribution to *Reviews of Modern Physics* from 1957, based on his presentation at the famous Chapel Hill conference [DR11], Deser [Des57] presented an argument according to which a *quantum theory of general relativity* could help remove the divergences that had plagued field theories for the last generation, by imposing a limit on the energy density of any material body or wave packet, independently of the specific fields involved:

A quantum field theory including gravitational effects sees to rid itself of point-singularity difficulties which neither part alone can avoid; classical relativity did tolerate point singu- larities in its solutions, which this more complete framework has no place for. Nonsingular matter distribution and nonsingular metric go hand in hand. This can be ascribed partly to the intimate coupling of geometry and kinetic energy, leading to limitations on energy concentrations, partly to the indefiniteness of the line element: the "smearing of the light cone" due to the appearance on an equal footing with Euclidean space of all other suitable Riemann spaces.

([Des57], p. 422)

Similar to Bronstein's emphasis on the difference between gravity and electromag- netism, Deser lists several crucial features of general relativity that distinguish it from other field theories, and may help escape the non-renormalizability thereof. In particular, it was his conviction that the (non-linear) characteristics of the gravita- tional coupling would allow one to evade the singularities that appear in other field interactions. This coupling, along with Bronstein's argument for a gravitational lower bound on measurability of field quantities in arbitrarily small spacetime regions (e.g., [Des57], p. 421, fn. 16), yield the characteristic (and in this case also minimal) length of quantum theory of general relativity of the order of 10^{-32} cm. Deser notes (but does not elaborate) that despite its smallness, this length may have observable consequences in lower energy regimes.

We learn in more detail about the quantum origins of this limit from a paper by Salecker and Wigner [SW58], which was also presented at Chapel Hill. The authors examined the limitation which quantum mechanics imposes on the possibility of measuring distances between events in spacetime, and argued that only an answer to this question could decide whether the gravitational field of atomic systems and of elementary particles is observable *in principle*. Note that it is necessary to answer this question *before* engaging in any discussion (of the sort we shall encounter below in Section 5.6) on the practical observability (or lack thereof) of gravitational effects at the atomic scale.

The starting point of the investigation is a physical clock whose purpose is to measure the spacetime interval between any two events consisting of collisions between material objects and light quanta. If such a clock existed with arbitrary accuracy, and if the recoil of light signals could be disregarded, it would be possi- ble to measure the interval with arbitrary accuracy. However, quantum uncertainty constrains the frequency of the clock, as well as its minimum mass. The uncer- tainty in energy (which constrains the frequency τ) is $\epsilon = 2\pi\hbar/\tau$, and the related minimum mass is $M > (\hbar/c^2\tau)(T/\tau)$ where T is the measurement duration. The latter constraint means that the wave packet of the center of mass of the clock will be confined, throughout the measurement interval time T, to a region of the size $c\tau$.

Neither of these two conditions requires that the physical dimensions of the clock will be limited. Nevertheless, the energy levels of the clock under consideration are extremely closely spaced, and this result, along with the requirement that the clock's collision with the light quanta will be strong enough to allow a reading of the measurement, indicate that the requirement of a clock with a small extension is non-trivial: in addition to the upper bound on the mass of the clock that comes from general relativity, this mass and its spread (uncertainty) will depend on the accuracy with which the interval is measured, the running time of the clock, and its size. In every case the smallest time interval which can be measured is roughly

$$t_{\min} \approx \frac{\hbar}{mc^2}. \tag{5.11}$$

That is, the time associated with the mass of one of the elementary systems out of which the clock is constructed represents an absolute minimum for measurable time intervals (see [DR11], pp. 176–177).

Salecker and Wigner then construct a possible microscopic clock and demonstrate the applicability of the above restrictions. For existing elementary systems, $m \approx 10^{-24}$ g, and so the corresponding time $t_{\min} \approx 10^{-23}$ seconds may be regarded as an absolute lower limit to measurable time intervals. Applying these considerations to measurements of the gravitational field of a particle of protonic mass, using a particle of electronic mass as a test particle, and choosing the most favorable clock model to serve as measuring apparatus, Salecker and Wigner find the uncertainty in the resultant measurement of the mass producing the field to be $\Delta M \approx 10^{-24}$ g. In other words, the uncertainty is of the same order as the mass itself. From this, they conclude that the gravitational mass of a single proton is not strictly an observable quantity.

The upshot of these considerations is similar to those of Bronstein's from a generation earlier: the Schwarzschild radius imposes a lower bound on the spatial volume in the measurability of field quantities – introduced in Bohr and Rosenfeld's paper [BR33] only as a finite but unbounded limit (relative to the particle size) – turning it into a *minimal* volume. On the other hand, the uncertainty associated with the measurements of the gravitational field in that volume justified the so called "fluctuations of the metric."

5.5.2 Alden Mead walks the plan(c)k

An argument that aims to establish the physical equivalence of these two limitations – namely, that a fundamental length as a limit on spatial resolution leads to uncertainties in the measurement of the gravitational field, and that, vice versa, uncertainties in the value of the gravitational field lead to a fundamental length – was conceived at the same time, but was only published several years later in two consecutive papers [Mea64, Mea66], by C. Alden Mead, an American professor of physical chemistry from the University of Minnesota. Apart from the inherent relevance of these two papers to our story, the bumpy road to their publication sheds some light on the reluctance to accept ideas about fundamental length in those days.

Mead, in a personal interview conducted with the author in Savannah (GA) in the fall of 2009, recalled that he started thinking about the dual role of gravity and the uncertainty principle already in 1958, when he was spending a year as a postdoc at Brookhaven laboratory.

What prompted this idea, says Mead, was actually a mistaken reading on his part of one of the many thought experiments that were elaborated during the Bohr–Einstein debate on the consistency of quantum theory, in which Bohr evades Einstein's harsh criticism by invoking the gravitational effects on the measuring clocks (for a description of this *Gedankenexperiment* see, e.g., [Ros66], pp. 599–603; for a criticism see [Shi00] and [Kie07], pp. 12–13). His (wrong) idea that gravity is necessary for the consistency of quantum theory was soon superseded by the idea that when considering gravitational effects in a quantum mechanical setting, one must do so consistently. Mead played a bit with the idea of trying to quantize gravity, but after a short while converged on a much simpler task, namely, the generalization of Heisenberg's microscope to include gravitational effects. The paper was sent to the *Annals of Physics* and then to *Il Nuovo Cimento*, but was quickly rejected by both journals with no feedback.[14]

Mead was stubborn enough to send the paper elsewhere, this time to the *Physical Review*, where it was also rejected but now with some comments from the referees, who seemed to take issue already with its first equation. Embarrassingly for the *Physical Review*, this equation was rather trivial. Mead tabled the paper and concentrated on publishing on other topics, but he returned to it in 1963. Mead wrote a short paper that proved the first equation to which the earlier referees objected, and sent it to the *Physical Review*. The referee report came quickly, recommending rejection since this was an obvious result. Mead then wrote to the editors, exposing the inconsistency in their earlier decision. His original paper was soon accepted, and appeared in print [Mea64] almost six years after it had been conceived.

With the clarity of hindsight, Mead's reasoning may seem almost natural to current quantum gravity practitioners, as it combines almost all the ingredients we

---

[14] According to Mead, presumably the editors saw an unknown author from a physical chemistry department and gave little credibility to its content.

have been discussing in the last two chapters: the operational and the relational perspectives from which a fundamental length is treated as a limitation on spatial resolution, its being motivated by the wish to avoid divergences in field theory, and its role either as characteristic (dependent on the mass of the relevant elementary particle) or as minimal (in the case of the inclusion of gravitational effects).

The same relational motivation led Mead to consider the relation between fundamental length and gravitational field fluctuations, and to see the limitation on spatial resolution as a limitation on the ability to realize coordinate systems. If a coordinate system is modeled as a distribution of fixed bodies and synchronized clocks in a given region of space, the limitation on the ability to realize it becomes a limitation on the accuracy with which one can synchronize the clocks. From a general relativistic perspective, however, the fluctuation in time of arrival (of the light signals used for the said synchronization) can therefore be regarded as a fluctuation in the spacetime metric. Note that the fluctuation results from uncertainty that is manifest in repeated measurements and in the non-reproducibility of the results, so that successive measurements give results that fluctuate around some mean value, in contrast to the notion of "absolute uncertainty" that was entertained in the 1920s (Section 5.2.1), and which accepts an uncertainty even in a single measurement by considering the impact of the system measured on the measuring apparatus.

It takes five equations and simple physical reasoning ([Mea64], p. B852) that combines Heisenberg's uncertainty principle and the gravitational limit on mass density to arrive at the conclusion that a particle cannot be localized by any repeated experiment to a region whose radius (in the natural units where both c and $\hbar$ are set to 1) is smaller in order of magnitude than $\sqrt{G}$ (where G is Newton's gravitational constant).

To demonstrate the applicability of this limit Mead suggested the following thought experiment. Suppose we want to measure the position of a free particle using a photon of frequency v, scattered by the particle into the aperture of a microscope, where it is focused and observed. We represent the direction of the trajectory of the photon from the particle to the microscope as a spread over an angle ϵ, and the radius in which the interaction of the photon with the particle is strong enough as r. These two parameters allow us to express the uncertainty in the measurement of the position coordinate of the particle x. The first arises from the limit on the resolution power of the microscope:

$$\Delta x \gtrsim \frac{1}{v \sin \epsilon} \gtrsim \frac{1}{v}. \tag{5.12}$$

The second arises from the fact the photon can be scattered from any point in the region of radius r surrounding the particle:

$$\Delta x \gtrsim r, \tag{5.13}$$

and since the photon cannot be focused during the time it interacts with the particle, a time τ must elapse between the scattering event and the recording of the measurement, that in general must be of the order of the time required by the photon to move the distance r from the particle or, in symbols, $\tau \gtrsim r$.

Mead first offers a non-relativistic analysis: the photon carries with it a gravitational field, and so during the time of its flight from the particle to the microscope, the particle experiences acceleration in the direction of the photon of the order

$$a \sim Gv/r^2. \tag{5.14}$$

If the particle does not attain relativistic velocities, and is originally at rest, it attains a velocity v in the direction of the photon which is proportional to the time required for the photon to escape the region r of strong interaction:

$$v \sim Gv/r, \tag{5.15}$$

and the distance it moves in that time r is

$$L \sim Gv. \tag{5.16}$$

Now, this motion of the particle is in the direction taken by the photon and is unknown, hence its projection on the x axis $L \sin \epsilon$ yields yet another uncertainty in the x-measurement

$$\Delta x \gtrsim Gv \sin \epsilon. \tag{5.17}$$

Combining (5.12) with (5.17) one obtains

$$\Delta x \gtrsim \sqrt{G}. \tag{5.18}$$

Making this analysis more precise by considering further the conservation of momentum in the scattering event yields an uncertainty in position due to the gravitational effects similar to (5.17), and hence does not change the final result.

The basic idea behind this result is very simple: in order to reduce the uncertainty in (5.12) one requires a photon with very high energy, but such a photon carries a very strong gravitational field, which in turns limits the ability to predict the future position of the particle. This latter effect is independent of the nature of the particle – a property characteristic of no other force but gravity and the principle of equivalence – and this is the reason for the difference between GTR and QED to which Bronstein had alluded, namely, that contrary to the case of the electromagnetic field, where the uncertainty can be mitigated by using a heavier particle, (5.18) is independent of mass, and thus represents a fundamental limitation on the possibility of localizing any conceivable particle.

The next step is to generalize this result to the relativistic domain: the uncertainty due to (5.17) becomes comparable to that due to (5.12) and (5.13) at about the same values of v and r where the velocity in (5.15) reaches the order of unity, and the Newtonian law of attraction loses its validity for a gravitational field which is changing rapidly in time, as is the case with the photon. Nevertheless, Mead shows that the result (5.18) still holds even in the fully relativistic case.

Mead then turns to discuss other methods to measure position, but his analysis shows that since they all involve objects that "take up space," they all yield the same order of magnitude as (5.18). He also demonstrates that the same type of uncertainty $\Delta T \gtrsim \sqrt{G}$ (where T is the clock reading) applies for the synchronization of two clocks using light signals. Here, again, the result holds irrespective of the mass or size of the clocks, and applies to *consecutive* measurements.[15]

The final part of the paper proves the equivalence of the notion of a fundamental length and the fluctuations of the gravitational field, when the latter is expressed in terms of uncertainty in measuring components of the gravitational field. This equivalence is characteristic of *gravitational* field measurements, as no other field has the property of imparting the same acceleration to all bodies irrespective of their mass or charge. It also allows one to constrain the regime in which the fundamental length can be postulated: if one wanted to introduce it in, say, the nuclear regime, the consequence would be a gravitational energy fluctuation greater than the Coulomb energy in the region of the nucleon, an effect which has not been observed.

The Mössbauer effect

A sequel to the paper [Mea66] was written during sabbatical leave at Birkbeck college which Mead spent with David Bohm's group.[16] Mead recalls that even after the paper and its sequel were published, there was little interest in the ideas they expressed, or in the suggestion to test them. Years later, prompted by a remark made by Frank Wilczek in *Physics Today* about the ubiquity of the notion of fundamental length in modern physics, Mead, by then retired, was struck by the discrepancy between this ubiquity and his experience, and wrote back to the journal [Mea01] recounting his difficulties in discussing ideas on fundamental length in the 1950s and 1960s.

In the second paper Mead discusses possible experimental set-ups that can probe the notion of fundamental length, due to its predicted effect on the broadening of spectral lines in the (then) newly discovered phenomenon of "recoilless emission,"

---

[15] Note, again, that one can measure T with arbitrary accuracy, but one cannot prepare a pair of clocks which will remain reliably synchronized over a finite period of time [AB61]. This means that setting up a Lorentzian frame by means of clock synchronization can only be done with a mean error of the order of the fundamental length.

[16] A young mathematical physicist in that group, Jeff Bub, returned with Mead to the University of Minnesota's physical chemistry department, but soon became drawn to the philosophy of quantum mechanics, and in the following years became one of the world's leading philosophers of physics. Bub would one day train the late Itamar Pitowsky, who was one of the author's academic mentors.

known as the Mössbauer effect [Mos62]. This effect involves photon emission from a nucleus bound to a crystal in which there is no kinetic energy transfer (hence no "recoil") to the internal degrees of freedom of the lattice – although, and this fact is sometimes glossed over, there is always energy transfer, hence recoil, to the lattice *as a whole*, i.e., to its center of mass [Eyg65].

Since the discovery of the Mössbauer effect, a lot of data have accumulated on frequency transitions in gamma emission, and Mead suggests that these data can constrain the hypothesis of fundamental length, as the uncertainty in the gravitational potential in the region occupied by the emitter leads to a broadening of that frequency. Thus, for example, a fundamental length of the order of 10^{-13} cm – the order of the radius of the nucleus – leads to a prediction in which the transition frequency ought to be very large, of the same order of magnitude as the average frequency, and is many orders of magnitude above the actually observed frequency. But a fundamental length of the order of $\sqrt{G}$ – the one suggested by Mead in his first paper on the subject – puts the broadening width just below the regime of the narrowest Mössbauer effect observed in the 1960s. To falsify this prediction one would have to measure gamma emission with a lifetime of several seconds or more, and exclude all other sources of broadening. Mead then suggests a possible experimental set-up that could achieve this feat ([Mea66], pp. 998–1002). As far as he knows, however, his suggestion was never taken seriously.[17]

5.6 Quantizing gravity – the philosophical debate

In this chapter we have seen how the notion of fundamental length motivated the attempts to unify quantum theory with the general theory of relativity in the 1930s through the 1950s. Most of the arguments we have presented rely on thought experiments that rest on some version or another of the so called semiclassical "disturbance" view of the uncertainty principle. This is evident in the attempts to generalize Bohr and Rosenfeld's [BR33] measurability argument to include gravity (Bronstein, Klein, Deser), and in the use of Heisenberg's microscope set-up (Mead).

The "disturbance" view has been criticized on many occasions as inadequate (see, e.g., [BR81]), and its implications with respect to the question of quantization of classical fields are commonly regarded as inconclusive [Ros63]. The reason for the inadequacy has to do with the fact that the "disturbance" view can be enlisted in support of an epistemic account of QM, which is hard to maintain: it suggests that there are definite, classical values, but that the dynamical limits of experiment (or the quantum nature of the measurement process) prevent their measurement.

---

[17] I found only two references that relate the Mössbauer effect to the idea of fundamental length after Mead. The first [AM73] mentions Mead; the second, a suggestion due to D. K. Ross [Ros84] (Section 2.5), was made independently of Mead's papers, and without any knowledge thereof.

However, if the "disturbance" view is suspect, or so the story goes, then so is the idea that it implies the necessity of the quantization of classical fields such as gravity.

In this section I shall first offer an alternative to the "disturbance" view that may help dispel much of the criticism of the thought experiments discussed above and their alleged support of hidden variables. I shall then suggest a more charitable reading of these thought experiments, that does not require one to see them as entailing in any way the quantization of the gravitational field.

5.6.1 Operationalism

There are two ways to blunt the criticism marshaled against the Heisenberg-microscope-type thought experiments we have been discussing. The first is conservative, the second is more radical, and both replace the "disturbance" view of the uncertainty principle with another, less problematic one.

The conservative alternative is also the simpler: one need only be reminded of the operationalist philosophy behind the thought experiments that motivated the introduction of fundamental length as a limitation on spatial resolution in measurements of field quantities. According to this operationalism, measurement outcomes are taken to be unanalyzable primitives, so one can remain agnostic about the question of whether a physical system possesses a definite value for a given property outside the context of measurement.

The twenty-first century version of this operationalism comes in two flavors. The first, subjective version [Bub00, BP10], interprets quantum probabilities as designating Bayesian degrees of belief, and not as designating ignorance of existing unknown values. Since the wave function designates our knowledge of the experimental result, the issue of hidden variables need not arise. The second, objective version [HS11], was mentioned in Section 3.2.4. It replaces the subjective view by interpreting quantum probabilities physically as objective measurement errors that arise from limitations on measurement resolution. Here also, the question of hidden variables need not arise.

The reason for the irrelevance of hidden variables is the following. On both operationalist accounts, the subjective and the objective, what matters is the state space of measurement outcomes, and not the metaphysical interpretation of the quantum state. That such an operationalist view (which elevates finite–resolution measurement outcomes into the status of the basic building blocks of the theory) is agnostic of metaphysics is by no means an argument against it, as long as it succeeds in reproducing from these building blocks, under certain conditions, the probabilistic predictions of QM and the structure of the Hilbert space that encodes them. Since on this view the only difference between classical and quantum probabilities resides in the structure of the probability space (Boolean versus non-Boolean), and not in metaphysics (a system having or not having definite properties when not measured), the question of hidden variables becomes practically moot. In other words, on the operationalist view defended here, interference phenomena are seen as evidence not for some metaphysical difference between classical and quantum mechanics ("is the moon there when nobody looks?"), but as evidence for a quantitative difference in *measure* employed in the respective probability theories, quantum probabilities being non-Boolean – represented as they are mathematically as subspaces of the Hilbert space and not as subsets of phase space [Pit94].

The subjective operationalist takes the Hilbert space structure for granted, and interprets it as a probability space – that one can do so in Hilbert spaces with dimension larger than 2 is a non-trivial fact that follows from Gleason's theorem [Pit89] – hence one avails oneself of all of its mathematical features, features that are part and parcel of its being a vector space over the complex *reals*. The objective operationalist, in contrast, is committed to finitism, and cannot enjoy such a luxury, and therefore must also show how the continuum of the Hilbert space can be approximated (along with the Born rule associated with its inner product) with a more fundamental structure whose basic building blocks are finite measurement results. As we have seen in previous chapters, currently there are two suggestions how to achieve this goal. The first involves constructing a vector space over the complex integers (see [HOST13a] and Section 2.4); the second involves reconstructing the space of possible dynamical transitions between any two physical states as a probability space (see [HS11] and Section 3.2.4).

In sum, although one could interpret the thought experiments that rely on Heisenberg's microscope as supporting the hidden variables picture of QM, this interpretation is just a matter of taste. If, in contrast, one chooses to remain agnostic about the state of a physical system outside the context of measurement, then the thought experiments we have been discussing begin to look rather benign: they do not attempt to analyze the measurement process, instead they focus on its observable outcomes. The uncertainty they represent results not from ignorance of an existing value, but from the requirement that consecutive measurements on the composed system (e.g., the particle and the photon) do not commute, so that the probabilistic predictions they yield obey the probability structure of QM.

5.6.2 Uncertainty from discreteness

We now step out of the comfort zone of agnosticism and look for more radical alternatives. Conservative readers who are averse to speculation may want to skip the next part and go straight to Section 5.6.4. Those who wish to stay, do so at their own risk.

The more radical alternative I suggest turns the table upside down; instead of accepting the uncertainty principle as a given, i.e., as a mathematical theorem of quantum theory, I propose to ask where it comes from. Answering this question, one is trying to underpin physically (and not only to characterize mathematically) the non-Boolean structure of quantum probabilities. This structure, recall, is manifest in non-commutativity, the uncertainty relations, and interference phenomena, and so the task is to show how these features can arise from assumptions about fundamental limitations on spatial resolution.

The intuition behind this physical underpinning was, as a matter of historical fact, already voiced by both Dirac and Heisenberg as early as the fall of 1926. Prompted by their attempts to establish the consistency of Heisenberg's matrix mechanics with the experimental data, data which Schrödinger's wave mechanics represented by means of continuously evolving causal processes in space and time, Dirac ([Dir27], pp. 622–624) and Heisenberg were concerned with the question of whether the formalism of the newly constructed quantum theory allowed for the position of a particle and its velocity to be determinable in a given moment in time:

> If it ends up that space–time is somehow discontinuous, then it will really be very satisfying that it wouldn't make sense to talk about, e.g., velocity $\dot{x}$ [the first derivative of position – AH] at a certain point x. Because in order to define velocity, one needs at least 2 points, which, in a discontinuum-world, just cannot lie infinitely close [benachbart].
> *(Heisenberg to Pauli on November 15, 1926 in [Pau79], pp. 354–355)*

Their negative conclusion, enforced by intuitions they shared about spatial discreteness and non-commutative geometry, has led to the common view of the uncertainty principle as a restrictive empirical principle on the precision of measurements, and to an alternative to Born's statistical interpretation of the state vector [Bor26] or to Jordan's version of quantum theory (both of which kept open the possibility that Nature is intrinsically indeterministic [DJ13]) as yet another context where probabilities enter into the formalism of QM, now as transition probabilities.

Holding as they did to the belief that Nature's dynamics is deterministic, and contrary to Born's statistical interpretation which identified the square of the amplitude of a *stationary* state vector with the probability that the system is in that state, the idea, shared both by Dirac ([Dir27], p. 641) and by Heisenberg ([Hei27], p. 62), was that the statistical element of QM arises only when experiments, or observations, are made:

> One can, like Jordan, say that the laws of nature are statistical. But one can also, and that to me seems considerably more profound, say with Dirac that all statistics are brought in only through our experiments. That we do not know at which position the electron will be the moment of our experiment, is, in a manner of speaking, only because we do not know

the phases, if we do know the energy ... and in *this* respect the classical theory would be no different. That we *cannot* come to know the phases without ... destroying the atom is characteristic of quantum mechanics.

(Heisenberg to Pauli, February 23, 1927; in [Pau79] Doc. 154, p. 377;
emphasis in the original)

This idea is also manifest in Heisenberg's remark that "the uncertainty relation doesn't refer to the past" ([Hei30a], p. 20), that is, that the uncertainty relations should be interpreted as the inability to measure both the *current* position and the *future* momentum of a particle. As we have already noted, the problem with this view was that, without an objective alternative thereto, the transition probabilities it involves were commonly interpreted as purely subjective and epistemic, representing as they do the lack of knowledge of the observer, whose measurement "disturbs" a pre-existing value (this criticism seems to have been raised for the first time in [vL55], see [Jam74], p. 73). We have just seen that the Bayesian can rectify this lacuna by recasting these transitions as mental updating, and – more relevant to our story – that the finitist can do so by treating these transitions as deterministic objective chances.

Heisenberg's obsession with fundamental length was discussed in Chapter 4. Dirac was less explicit, but it is by now well documented that his work was influenced by ideas consistent with the hypothesis of finite spatial resolution, for example, finitist and operational ideas such as Whitehead's method of extensive abstraction, or the geometrical interpretation of non-commutativity [Dar92]. The latter was for Dirac the most distinctive feature of the new quantum theory that separated it from the classical theory (e.g., [Dir29]). Dirac was also well aware of the methodological problems that accompanied the notion of a finite spatial resolution (e.g., the loss of relativistic causal order at short distances), but was not as worried about these problems as his contemporaries were. This attitude is evident in his solution to the problem of the self-energy of the classical electron, which introduces a finite electron radius [Dir38], in his interpretation of this cutoff as an inherent feature of spacetime (and not just as a limit on the theory's applicability), and in his famous and persistent criticism of the renormalization method [Dir63].

My suggestion for the physical underpinning of the uncertainty relations follows these intuitions, and rests on two simple assumptions: (a) it is *in principle* impossible to perform a position measurement with infinite precision; (b) every measurement can be reduced to (or construed as) position measurement.

Assumption (a) follows from a finitist view of nature (see, e.g., [Fey65], p. 57 or [Fre90], p. 255). Note that, as a matter of fact, assumption (a) is consistent with any position measurement we have ever made. That assumption (b) is plausible follows from the fact that a well-known theory, namely Bohmian mechanics, is empirically indistinguishable from non-relativistic QM. Relying on a stronger version of assumption (b) as it does, namely, that every physical property ontologically supervenes on position,[18] Bohmian mechanics nevertheless reproduces the Born rule under an initial equilibrium assumption (i.e., a uniform probability distribution in position space). Consequently, any counterexample to assumption (b) would automatically be a counterexample to the applicability of non-relativistic QM. Indeed, in his path integral formulation of QM, Feynman famously states that "a theory formulated in terms of position measurements is complete enough in principle to describe all phenomena" ([FH65], p. 96). John Stewart Bell is yet another physicist who promoted assumption (b) on many occasions (e.g., [Bel87], pp. 52–62).

As an example of assumption (b) consider the following model of momentum measurement in one dimension, which involves measuring momentum by measuring spatial displacement: by measuring how far a particle has traveled from its original position x (assuming no forces) after the time $t = T$, one can decide its momentum; if the position displacement is $x + \Delta x$, then the velocity is $\Delta x / T$ and the momentum is $p = m \Delta x / T$. This simple model nicely exemplifies the idea that the two measurements are position measurements whose accuracies are inversely proportional (see Figure 5.1).[19] It also demonstrates that – in accordance with the agnostic operationalism defended above – what is being "disturbed" is not the state of the particle, but the accuracy of the consecutive measurement.

To see how these two assumptions physically underpin the uncertainty relation let us take the mathematical formulation thereof which dispenses with standard deviation as an inadequate measure of spread of the probability distribution, and replaces it with more appropriate measures of spread (see [HU83, UH85, HU88b]). Consistent with Heisnberg's intuition above, in this formulation the uncertainty relation represents the inability to measure precisely both the *current* position and the *future* momentum of a particle. In the context of Heisenberg's microscope thought experiment it can be written as ([HU90], p. 133):

$$w_x W_p \geq C, \tag{5.19}$$

where C is a constant of order unity. Relation (5.19) connects the resolving power of the measuring apparatus (w_x) with the predictability of the momentum of the

---

[18] Here I adopt only the weaker, operationalist version of this assumption, while remaining agnostic about the ontology of properties other than position.

[19] Let δx be the inaccuracy of a position measurement, i.e., the error in precision of the position measurement of a particle which is either at rest or in motion. In order to resolve smaller distances we need higher energy, and this energy translates, in the absence of other forces, into work, i.e., to an additional distance ΔX that the particle may travel in a given time $t = T$ over and above the distance X it would have traveled at time $t = T$ if no position measurement were performed. This means that $\Delta X \sim (\delta x)^{-1}$. But ΔX is also the error δp in the spatial measurement of the momentum p and so $\delta p \sim (\delta x)^{-1}$.

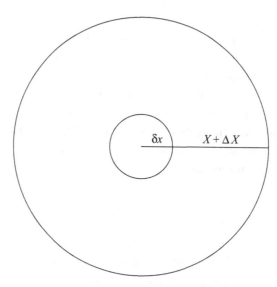

δx $X + \Delta X$

Figure 5.1 Momentum measurement as position measurement.

measured object after the measurement (W_p). Mathematically, the former magnitude is the width of the wave function of the state of the object in position space, that represents how far that state is from a spatially translated copy of itself [HU89]. The latter magnitude is the overall width of the wave function of the state in momentum space; it represents how narrow is the probability distribution of momentum, i.e., how small is the inaccuracy in its prediction.

By assumption (a) the first magnitude (w_x) is finite; by assumptions (b) and (a) – since we actually measure position when we measure momentum – the second magnitude (W_p) can be construed as representing inaccuracy in position measurement and hence is also finite. We can physically justify relation (5.19) once we realize that the inaccuracy of the momentum measurement depends on the (finite) inaccuracy of the earlier position measurement, and vice versa (see Figure 5.1). Note that, in accordance with our goal to propose an objective notion of uncertainty and probability, this underpinning also replaces Heisenberg's intuition about uncertainty arising from "disturbance" – in this case the disturbance of the momentum of the particle by the position measurement – with an objective, "hidden-variable-free" notion of uncertainty that arises from an inherent limitation on spatial resolution: the point is that nowhere does this view refer to the actual (but presumably unknown) state of the particle; what is being "disturbed," or modified, is the consecutive measurement, whose inaccuracy is inversely proportional to the inaccuracy of its alternate measurement.

So much for non-commutativity. What about interference? For this, consider the proverbial double slit experiment, where interference is explicitly addressed.

Bohr famously argued ([Boh83], p. 25) that the attempt to discern which slit the particle goes through "destroys" the interference pattern on the photographic plate. If one considers Bohr's set-up [HU90], one notes that the detection of the slit the particle goes through relies on the recoil of the slitted screen due to the passage of the particle in that slit. This recoil sets an upper bound on the initial knowledge of the momentum of the movable screen: for this recoil to be distinguishable, the initial momentum of the screen along the y-direction (the direction perpendicular to the flight of the particle) must be known with an uncertainty bounded from above by the same momentum recoil.

Next, Bohr reasoned that this upper bound in momentum imposes, via the uncertainty relation, a lower bound on the overall width of the wave function in position space, which is greater than the width of the interference band, and that if one attempts to detect the path of the particle, the interference vanishes. While Bohr's reasoning was incorrect,[20] his conclusion was nevertheless, right! This conclusion, namely, that the conditions under which a measurement that allows distinction between the two paths the particle could have taken exclude the conditions under which interference can be observed, is demonstrated succinctly in [HU88a] and [HU90], pp. 135–136, where the appropriate width (or measurement inaccuracy) which relates the interference pattern to the distinguishability of the two momentum states of a movable screen (associated with the passage of the particle through one slit or the other) is of the *visibility* of the interference pattern (rather than the width of the interference bands that Bohr used in his reasoning). Thus, whenever these states can be completely distinguished, that is, if the slit through which the particles pass can be determined with certainty, the visibility of the interference pattern also vanishes. For an actual (and not only *Gedanken*) experiment of precisely this sort see [BSH+98].

Given assumption (b), both widths, the distinguishability of the momenta and the visibility of the interference pattern, physically represent inaccuracies of spatial displacement measurements. That a momentum measurement can be construed as a spatial displacement (or distance) measurement was argued above. That a measurement of the visibility of the interference pattern can be so construed follows from a simple analysis of interferometry: there, *visibility* is defined as a function of the *intensity* of the beams which, in turn, is a function of the *spatial displacement* of the mirrors that constitute the interferometer [Uff85]. As shown in [HU90], the two resolutions are dependent since the visibility is proportional to the matrix element which is a direct measure of the distinguishability of the momentum states of the screen. In our reconstruction of the momentum measurement as position

---

[20] The reasoning involved an inappropriate measure of probability spread, namely the translation width of the screen in position space, which turns out to be insensitive to the width of the interference pattern, see [HU90], p. 131.

measurement (Figure 5.1), this situation occurs when ΔX is bounded from above and is equal to δx, yet the conditions for observing interference require a spatial accuracy $d < \delta x$.

5.6.3 Though shalt not commute!

I have suggested that non-commutativity arises from fundamental limitations on spatial resolution, but is not this physical underpinning too strong? Indeed, assumption (a) may at first look far too radical, as its combination with (b) seems to make any pair of consecutive measurements (and not only incompatible ones) strictly non-commuting. The key, however, to the reconciliation of the above physical underpinning of relation (5.19) with classical mechanics (where measurements always commute), or with the case of commuting operators in QM, i.e., where there are no interference terms, lies in the following operational interpretation of commutativity as a manifestation of the overdescription, or the coarse graining, of the discrete.

First note that the overdescription of the discrete, discussed in Sections 2.1 and 3.2.2, is germane to all our physical theories, QM included. These theories employ a mathematical machinery that has much more structure than is required to model physics. Consequently, a lot of effort is spent in disabling or reinterpreting these redundancies, so that the modeling can be done in spite of them.

I suggest to view the notion of commutativity as yet another example of such an overdescription. In classical mechanics commutativity represents the coarse grained character of classical spatial resolution, which is presumably infinite. Spatial measurements can be regarded, from a coarse grained perspective, *as if* they are done with infinite resolution, and consequently one can make the inaccuracies thereof arbitrarily small. Mathematically this means that classical measurement errors, while possible, are just a practical nuisance that can be compensated for and eliminated by arbitrarily enlarging the scope of the physical description without cost. Operationally, commutativity ensues from the fact that infinite precision is allowed so fundamentally there is no bound on physical resources. Consecutive measurements in classical mechanics thus appear to be (and so can be described as) independent of each other's inaccuracies *for all practical purposes.*

In QM the situation is no different. Here commutativity represents the operational independence of any two measurements, or their compatibility. In contrast to the uncertainty relations, or, say, to the case of any two consecutive orthogonal spin measurements, in those cases where measurements commute, the accuracy (or lack thereof) in one does not depend on the accuracy (or lack thereof) in the other. From the perspective developed here, this independence can arise in two types of scenarios. First, the physical process of construing, or performing, at least one of the measurements as a position measurement involves, as a matter of fact, correlations with more degrees of freedom, and leads, as in the classical case, to a similar spatial coarse graining, that makes the measurements *practically* insensitive to the respective inaccuracies (think, for example, of temperature measurement with an analog thermometer whose scale is too coarse to represent minute errors in position). Such coarse graining – commonly referred to as "decoherence" – makes the measurements practically insensitive to their respective inaccuracies. Second, a temporal coarse graining is also possible, where commutativity follows from an extremely (ideally infinite) long measurement process, as in the quantum adiabatic theorem ([Mes61], pp. 739–746), or in the so called "protective measurement" scenarios [AAV93].

In both types of coarse graining one can treat the inaccuracies in spatial measurements as if they are arbitrarily small, hence – *for all practical purposes* – one can describe the pair of consecutive measurements as commuting.

This reasoning suggests that while the above physical underpinning of the uncertainty relations may seem too radical, there is a way to make it consistent with the classical or the quantum formalism, and with the predictions of both theories, at least in the non-relativistic domain (the generalization to the relativistic domain may be more involved, but certainly not *a priori* impossible). The price, however, is that non-relativistic QM is now seen as a phenomenological, "effective" theory, whose mathematical structure (the Hilbert space), rather than being fundamental, is actually a mere set of ideal analytical tools for computing the probabilities of future states of an underlying deterministic and discrete process, from the (inherently) limited information we can have about that process.[21]

5.6.4 Quantization?

There is, of course, no need to accept such a radical view, and arguably more work is needed to make it more plausible. Returning to a less speculative mode, I am definitely not suggesting that the proprietors of the variety of thought experiments presented above were conscious in any way of the considerations of the last three sections; all I offer is an alternative to the "disturbance" view they are accused of relying on, that renders them less problematic. This leads me to the second step in their rehabilitation: one need not accept the Bayesian view of QM (or its objective, finitist, alternative for that matter) in order to appreciate another possible charitable reading thereof.

Such a charitable reading requires us to see these thought experiments not as supporting arguments for the necessity of the quantization of the gravitational field (as critics of the "disturbance" view suggest). Rather, we should see these thought experiments as supporting arguments that seek to establish the epistemic

---

[21] The implications of this view on the "interpretations" of QM are dire; they will be discussed elsewhere.

coherence, or the limits of applicability, of a putative theory of quantum gravity, *if* such a theory existed, gravity being what it is, namely, so different than any other interaction in that it couples to everything, irrespective of the mass under consideration. This reading blunts most of the thrust of these thought experiments, *if* they are intended to be construed as supporting the necessity to quantize gravity. I believe they should not be construed as such, and in what follows I will try to convince the reader that they also need not be construed as such, even if, as a matter of historical fact, they were so construed by many theoretical physicists.

The story here is well known. At the end of the 1950s and the beginning of the 1960s, a heated philosophical debate took place on the question of the quantization of the gravitational field, and in the context of this debate the notion of fundamental length assumed an interesting methodological role, as some saw it as giving support to the claim that the gravitational field must be quantized.

An argument of this sort was made by Peres and Rosen [PR60], who used the (by then standard) analogy between Bohr and Rosenfeld's paper on measurability of the electromagnetic field quantities in QED and the weak, quasi-static, case of the gravitational field, and arrived via considerations similar to those of Bronstein and Mead at the idea that this quantization leads to uncertainties in the values of the Christoffel symbols due to measurement limitation on the interaction of the field with massive test bodies. They concluded that these uncertainties in the measurement of the gravitational field imply the necessity of its quantization.

It was Rosenfeld himself, however, who questioned the analogy between his and Bohr's measurability limitation result from 1933 and the case of gravitational theory [Ros66]. For him, as long as an actually completed quantum theory of the gravitational field did not exist, the physical meaning of the critical quantities such as the Planck mass, the Planck length and the Planck time remained uncertain. Rosenfeld was also skeptical about the possibility of empirical evidence for quantum gravitational effects [Ros63], and, while not declaring them conclusive, presented many arguments to the effect that the gravitational field is a classical entity, that need not be quantized.

The lack of empirical motivation for quantization of the gravitational field was – and as we shall see below, still is – a real problem. Worse, several calculations showed that even if effects existed that could motivate quantization *in principle*, they would be too small to be observable *in practice*. Feynman [FMW03], for example, calculated that the gravitational field contribution to the ground state of a hydrogen atom would change the wave function phase by just 43 arcseconds after 100 times the age of the universe. Of course, this does not mean that all effects of this sort will be unobservable *in practice*; it just means that one should look for circumstances that are sensitive to such tiny effects, either in one's laboratory on Earth, or at astronomical scales [EMN99]. We shall discuss this in more detail in Chapter 8.

With empirical evidence absent, the debate ensued on the logical necessity (or lack thereof) of quantizing the gravitational field. The attitude Rosen and Peres expressed in their paper was prevalent in the late 1950s, as can be seen from the conversations that took place in the 1957 Chapel Hill conference:

GOLD said the theory of electrodynamics required quantization because there was a broad range of phenomena that had to be accounted for; for example, the infrared catastrophe had to be somehow explained. With gravitation, one could say that either it must be quantized because otherwise one might get into a contradiction with the logical structure of quantum theory, or else that there exists a broad range of phenomena which the classical theory (without quantization) is unable to explain. But this latter range of phenomena seems to be missing, unless there is gravitational radiation and some kind of "infrared" divergence occurs there. WHEELER said that it might turn out that the elementary particles depend on gravitational fluctuations for their stability. GOLD said that this was only a hope, and WHEELER agreed that a proof was missing. ANDERSON asked if there was anything wrong with the argument that if it were possible to measure all gravitational fields accurately, that one would have enough information to violate the uncertainty principle. WHEELER answered that this was Gold's first point. GOLD said that he wanted still to be convinced that one gets into a contradiction by not quantizing the gravitational field.

([DR11], pp. 243–244)

The idea that many shared was that the question of the quantization of the gravitational field was really a question about the universality of the quantum theory. Feynman, in that very conference, pursued this line to its extreme ([DR11], pp. 244–260), suggesting several thought experiments that today might be grouped under the title of "macroscopic interference," in which an initial quantum superposition is amplified to affect massive bodies and the gravitational field they carry. After some back and forth with Gold, Bondi, and Rosenfeld, Feynman admitted that quantization was not the only option; an action-at-a-distance theory of gravitation could be another possibility ([DR11], p. 259; see also Unruh [Unr84], p. 238), in which the uncertainty in measurement was a result of quantum theory alone, without gravitation, as, for example, Gold remarked:

It could be that inaccuracies are always introduced because no experiment can be, finally, gravitational only. It is possible that the existing quantum theory will already always make sure that nothing can be measured without sufficient accuracy.

([DR11], p. 258)

That the amplification of the quantum superposition is not forced on us is yet another conclusion of the critique of the putative implications of the "disturbance" view of the uncertainty principle we have mentioned above. Note, however, that under the charitable reading we have offered (according to which the notion of minimal length arises when one introduces gravity as a possible limitation on spatial measurements additional to the ones already resulting from the existing quantum theory of fields), the arguments for this additional, gravitational, limitation on measurement of field quantities need not be regarded at all as arguments for the necessity of the quantization of gravity; the gravitational constraint on spatial measurement could be seen just as one possible source for the limit of applicability of such a theory, if it existed.

The modification of Bohr and Rosenfeld's result to include gravity relies on the premise that general relativity imposes an additional limit on the mass density that can be used for localization of any test particle (see also Unruh [Unr84], pp. 241–242). The Bohr and Rosenfeld argument, however, does not entail that the electromagnetic field must be quantized; it can also be seen as delineating the limits of applicability of a putative theory of QED, if it existed. All it aims to establish, is that (*pace* Landau and Peierls [LP31]) within these limits of applicability, the putative theory of QED would be epistemically coherent, i.e., consistent with our measurement abilities in such a way that would not preclude its own verification.

For the same reason, Mead's argument for the equivalence between a fundamental length as a limit on spatial resolution and the uncertainties in the value of the gravitational field need not entail the quantization of gravity; it can also be seen as delineating the limits of applicability of a putative theory of quantum gravity, *if* such a theory existed. All it aims to establish, analogously, is that, within these limits of applicability, the putative theory of quantum gravity would be epistemically coherent, i.e., consistent with our measurement abilities in such a way that would not preclude its own verification.[22] This is precisely what Rosenfeld meant when he said that his argument with Bohr from 1933 only established "the consistency of the way in which the mathematical formalism of a theory embodying such quantization is linked with the classical concepts on which its use and analyzing the phenomena rests" ([Ros63], p. 354), and, presumably, is also what Mead meant when saying that he sought to introduce gravity into quantum theory in a consistent way, i.e., in a way that respects the intrinsic difference between gravity – with its universal coupling – and all other matter fields.

Under the charitable reading I offer here, and contrary to what Rosen and Peres have claimed, the notion of fundamental length that results from extending Bohr and Rosenfeld's considerations to the gravitational field is thus in no way an argument for the quantization of the latter. At most it shows what emerges if a unified theory of quantum field theoretic and gravitational interactions is constructed in a consistent, epistemically coherent way. This is also the way Mead regards the situation today.

---

[22] One could imagine an argument, analogous to that of Landau and Peierls, against the epistemic coherence of any putative theory of quantum gravity, that relies on the idea that single measurements of the components of the gravitational field cannot be made within this theory due to gravitational singularities, and concludes that the theory precludes its own verification. Mead's argument is thus a response to an imaginary attack of this sort.

Back at the Chapel Hill conference, Rosenfeld concluded the discussion on the necessity of the quantization of the gravitational field, by pushing it back to the question of empirical motivation:

It is difficult for me to imagine a quantized metric unless, of course, this quantization of the metric is related to the deep-seated limitations of the definitions of space and time in very small domains corresponding to internal structures of particles. That is one prospect we may consider. The whole trouble, of course, which raises all these doubts, is that we have too few experiments to decide things one way or the other.

([DR11], p. 259)

Several years later, in a reply to Belinfante [Inf64], he argued further against a fully quantum description of matter and gravitational fields, saying that Einstein's equations may merely be thermodynamical equations of state that break down for large fluctuations (a similar view was expressed in much more detail years later in [Jac95]), so that the gravitational field may only be an effective, and not a fundamental, field ([Kie07], p. 19).

By now the reader hopefully appreciates that from the perspective we have been developing in this monograph, any argument for the logical necessity of the quantization of the gravitational field is as wrong headed as the *a priori* arguments for the discreteness of space within the philosophy of mathematics. Our discussion in the first two chapters converged on the idea that whether or not a hybrid theory (discrete up to a certain level, continuous above it) exists cannot be decided as a matter of logic. The reason why most physicists working on quantum gravity would be reluctant to regard the semiclassical theory in which the gravitational tensor $G_{\mu\nu}$ (i.e., the spacetime geometry) couples to the *expectation values* of the stress–energy observable $\hat{T}_{\mu\nu}$ given that the quantum state of the matter field is Ψ,

$$G_{\mu\nu} = 8\pi \langle \hat{T}_{\mu\nu} \rangle_\Psi \tag{5.20}$$

(in natural units, i.e., when $G = \hbar = c = 1$) as fundamental, is not because of some no-go arguments (to be discussed in Chapter 8) but because this route has not proved to be productive. Since (5.20) only gives consistency conditions rather than a complete solution (typically many quantum states share any given expectation value), it requires going back and forth between the (classical) dynamics of the gravitational field and the (quantum) dynamics of the matter fields, and this back and forth involves, from a methodological point of view, "more negotiating pitfalls than producing positive results" ([HC99], p. 13).

To sum up, and contrary to the deeply entrenched folklore in the quantum gravity community, the notion of fundamental length, while arising naturally from the considerations of measurement limitation which involve both quantum field theory and general relativity, does not in any way imply that a unification of both theories requires as a matter of logic the quantization of the gravitational field. The only logical relation that does hold here is that when this unification includes a quantization of the gravitational field, it inevitably leads to imposing a fundamental length, whose meaning can be interpreted either ontologically or epistemically. Whether or not one should quantize the gravitational field remains an open question. As is always the case in physics, empirical evidence remains the final arbiter. To repeat Rosenfeld's immortal verdict, "even the legendary Chicago machine cannot deliver sausages if it is not supplied with hogs" ([Ros63], p. 356).

6

Einstein on the notion of length

6.1 Outline

Considerable attention has been drawn lately to the distinction, attributed to Einstein, between principle and constructive theories, and to the methodological importance this may have for scientific practice. Viewed as part of the context of discovery, however, this distinction is rarely acknowledged as having any philosophical significance. An exception is Howard [How04] who urges us to regard it as one of Einstein's most valuable contributions to twentieth century philosophy of science.

Admittedly, while Einstein was not the first to introduce the distinction between principle and constructive theories to theoretical physics, he definitely popularized it when reflecting on the conception of the special theory of relativity (STR henceforth). Expressing the novelty of the theory, Einstein ultimately chose the principle view over the constructive one, but his ambivalence with respect to this choice (and his misgivings about what he regarded as its unfortunate implications on the foundations of quantum mechanics) are well known [Sch49b, Jan00, Bro05a, Bro05b].

Also well documented is the attempt, made by Einstein's contemporaries Lorentz and FitzGerald, to think about the kinematical phenomena of electromagnetism in constructive – dynamical – terms (see, e.g., [Jan95]). Other physicists who expressed, along with Einstein (as some believe), dissenting constructive views of STR are less well known in this context. They include Weyl, Pauli, and Eddington in the 1920s, W. F. G. Swann in the 1930s and the 1940s, and L. Janossy and J. S. Bell in the 1970s.

In a recent monograph entitled *Physical Relativity*, Harvey Brown [Bro05b] adds his voice to this distinguished list of unconventional voices, arguing that the universal constraint on the dynamical laws that govern the nature of non-gravitational interactions, namely, their Lorentz covariance, is the true lesson of STR. My modest goal in this chapter is to examine one, presumably contentious, issue included in Brown's controversial view. This issue is purely historical, and concerns Einstein's attitude towards the constructive approach to STR. Brown ([Bro05b], pp. 113–114, [Bro05a]) suggests that Einstein's ambivalence with respect to his choice of the principle view of STR warrants annexing him to the constructive camp. Here I shall suggest an alternative interpretation. The ulterior motive behind this historical detour is that it also sheds new light on the way Einstein saw the relation between dynamics and geometry in spacetime physics, which, as it turns out, is highly relevant today in attempts to solve the quantum gravity problem.

In order to achieve this goal I shall examine a rarely cited correspondence between Einstein and the physicist W. F. G. Swann, mentioned only briefly in the vast literature on Einstein. In this correspondence Swann presents Einstein with his constructive approach to STR, wherein rods and clocks are not introduced as primitive building blocks, or as "independent objects," but are taken instead to be material bodies obeying the Lorentz covariant laws of the quantum theory of matter. Einstein, in response, argues cryptically that any dynamical formulation of STR must introduce a primitive notion of length.

In what follows I shall present the broad historical and philosophical context for Einstein's remark, and offer several possible interpretations of his letter. As we shall see, both the debate on the role of geometry in dynamical theories that purport to "derive" spacetime from more fundamental building blocks, and Einstein's insight in this regard, will prove instructive when we turn to explore current approaches to quantum gravity.

6.2 Constructing the principles

6.2.1 Einstein

In his famous letter to the London *Times* from November 28, 1919, Einstein mentions a distinction between two types of scientific theories, namely "constructive" and "principle" theories (I quote the second, third, and fourth paragraphs from [Ein54a]):

We can distinguish various kinds of theories in physics. Most of them are constructive. They attempt to build up a picture of the more complex phenomena out of the materials of the relatively simple formal scheme from which they start out. Thus the kinetic theory of gases seeks to reduce mechanical, thermal and diffusional processes to the movement of molecules, i.e., to build them up of the hypothesis of molecular motion. When we say that we have succeeded in understanding a group of natural processes, we invariably mean that a constructive theory has been found which covers the processes in question.

Along with this most important class of theories there exists a second which I will call 'principle theories'. These employ the analytic, not the synthetic method. The elements which form their basis and starting point are not hypothetically construed but empirically discovered ones, general characteristics of natural processes, principles that give rise to mathematically formulated criteria which the separate processes or the theoretical representations of them have to satisfy. Thus the science of thermodynamics seeks by analytical means to deduce necessary conditions which separate events have to satisfy, from the universally experienced fact that perpetual motion is impossible.

The advantages of the constructive theory are completeness, adaptability and clearness. Those of the principle theory are logical perfection and security of the foundations. The theory of relativity belongs to the latter class . . .

What does this distinction amount to? According to Einstein, in a principle theory such as thermodynamics, one starts from empirically observed general properties of phenomena, for example, the non-existence of perpetual motion machines, in order to infer general applicable results without making any assumptions about hypothetical constituents of the system at hand. Another example of a principle theory in which one employs "the analytic, not the synthetic method" is STR. Its building blocks – that velocity does not matter and that there is no overtaking of light by light in empty space – are "not hypothetically constructed but empirically discovered." Statistical mechanics (SM) and its predecessor the kinetic theory of gases, on the other hand, are constructive theories. They begin, says Einstein, with certain hypothetical elements and use these as building blocks in an attempt to construct models of more complex processes. Although ultimate understanding requires a constructive theory, Einstein admits in 1919, progress in theorizing is often impeded by premature attempts at developing constructive theories in the absence of sufficient constraints by means of which to narrow the range of possible constructions. It is the function of principle theories to provide such a constraint, and progress is often best achieved by focusing first on the establishment of these principles.

It is hard to overestimate the importance Einstein's historians attribute to the three short paragraphs quoted above.

Emphasizing Einstein's famous appreciation of the wide applicability of thermodynamics, Klein [Kle67] sees the distinction between principle and constructive theories as yet another indication of the inspirational power thermodynamics had on Einstein's thought, especially in conceiving STR. Janssen [Jan00] believes that the distinction epitomizes Einstein's ambivalence towards physics theorizing in general. "Einstein resorted to the 'principle' type of theory," says Janssen, "when he did not have a strong vision of what a satisfactory [constructive – AH] model might look like." "Since he saw this type of theorizing essentially as a physics of desperation," concludes Janssen, "his methodological pronouncements later in life promote the 'constructive' approach, which had never gotten him anywhere, rather than the 'principle' approach that had led to all his great successes." Howard [How04] sees Einstein's distinction as his most original contribution to twentieth century philosophy of science. "While the distinction first made its way into print in 1919," says Howard, "there is considerable evidence that it played an explicit role in Einstein's thinking much earlier."[1]

Howard argues further that in Einstein's hands the distinction between principle and constructive theories became a methodological tool of impressive scope and fertility. This point, while not appreciated as it should be in the philosophy of science community, is unproblematic. Einstein's attitude towards the constructive approach to theoretical physics in general and to STR in particular, on the contrary, is still under dispute. The peculiar ambivalence Einstein demonstrates in this respect leads Brown [Bro05b, Bro05a] to present him as a putative supporter of the constructive approach to STR.

6.2.2 *The constructive approach to STR*

Several authors [Bel92, Bro93] have suggested that the dawn of the constructive approach to STR can be traced back to a letter, written in 1889 by G. F. FitzGerald, to the remarkable English auto-didact, Oliver Heaviside, concerning a result the latter had just obtained in the field of Maxwellian electrodynamics. Some months later FitzGerald exploited the idea he had expressed in that letter, namely, that the distortions suffered by an electric field that surrounds a charged particle traveling through the ether may be applied to a theory of intermolecular forces, to explain the baffling null result of the Michelson–Morley experiment. In this note, a distinct precursor of the FitzGerald–Lorentz contraction, a cornerstone in the kinematic component of STR, appears for the first time.

As Brown puts it ([Bro05b], p. 2), following Einstein's brilliant 1905 work on the electrodynamics of moving bodies, and the geometrization of this work by Minkowski (which proved to be so important to the development of the general theory of relativity), "it became standard to view the FitzGerald–Lorentz [contraction – AH] hypothesis as the right idea based on the wrong reasoning." Brown strongly doubts that this standard view is correct, and in his monograph *Physical Relativity* he joins other physicists who expressed, along with Einstein himself, or so Brown claims, dissenting constructive views of STR.

Brown's aim is to advocate what he calls "the big principle" of the constructive view ([Bro05b], p. 147): that the universal constraint on the dynamical laws that govern the nature of non-gravitational interactions, namely, their Lorentz covariance, is the true lesson of STR. On this view, the explanatory arrow in STR between the structure of spacetime and the behavior of rods and clocks is reversed: if one

---

[1] Here Howard refers to Einstein's remark on Boltzmann's entropy principle, $S = k \log W$, which served as the constraint that suggested his own light quanta hypothesis (see also Einstein's letters to Ehernfest and Sommerfeld from 1907 and 1908, respectively, mentioned in ([Bro05a], p. S85)): "Boltzmann's magnificent idea is of significance for theoretical physics ... because it provides a heuristic principle whose range extends beyond the domain of validity of molecular mechanics" [Ein15], p. 262.

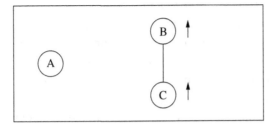

Figure 6.1 Bell's thread.

could achieve a dynamical underpinning of this behavior with an ultimate Lorentz covariant theory of matter, then the "mystery of mysteries" (i.e., how material bodies such as rods and clocks are supposed to know which spacetime they are immersed in and hence to contract and dilate accordingly)[2] will be dispelled, and Minkowski spacetime will regain its appropriate status as a "glorious non-entity" [BP06].

Bell's thread

The difference between the constructive and the principle approaches to STR, on final account, is an epistemological one, and as such can be represented in terms of explanatory strategies. Taking our cue from Brown, who emphasizes that one of his motivations for adapting the constructive approach is the reversal of the explanatory arrow in STR from geometry to dynamics, it is instructive to confront the principle and the constructive views in the context of Bell's thought experiment that appears in the opening paragraphs of his famous *Lorentzian Pedagogy* paper ([Bel87], pp. 67–68).[3]

Bell considers the following situation (see Figure 6.1): three small spaceships, A, B, and C, drift freely in a region of space remote from other matter, without rotation and without relative motion, with B and C equidistant from A. On reception of a signal from A the motors of B and C are ignited and they accelerate gently. Let B and C be identical, and have identical acceleration programs. Then (as reckoned by the observer A) they will have at every moment the same velocity and so remain displaced one from the other by a fixed distance.

Now suppose that one end of a taut thread is attached to the back of B and the other end to the front of C, assuming that the thread does not affect the motion of the spaceships. According to STR, the thread must Lorentz contract with respect to A because it has relative velocity with respect to A. However, since the spaceships maintain a constant distance apart with respect to A, the thread (which we have

---

[2] A rod immersed in a Minkowski spacetime is affected when set in motion while a rod immersed in Galilean spacetime is not.

[3] Bell mentions that credit to this thought experiment is due to [DB59].

assumed to be taut from the start) cannot contract; therefore a stress must form until at high enough velocities the thread finally reaches its elastic limit and breaks.

Bell mentions that the knee-jerk reaction of one of his colleagues as well as the consensus in CERN's theory division was that the thread will not break. Further reflection, however, reveals that despite the views that deprive Lorentz contraction of any reality, STR predicts that (1) the thread will break, and that (2) all the observers in this set-up, namely, A, B, and C, will agree on (1).[4] Setting aside the interesting sociological issues that the reactions to this thought experiment reveal, let us examine the two possible explanations one can give for facts (1) and (2).

According to the standard explanation to which the principle view of STR subscribes (see, e.g., [HT90], pp. 32–34), the absolute, invariant, distance measure in spacetime between two spatiotemporal events is the four-dimensional interval between them, and our spatial and temporal measurements of these events are nothing but covariant projections of this invariant measure on particular reference frames. Consequently, what explains the breaking of the thread is the fact that, since the spaceships are accelerating, this projection (of the absolute four-dimensional interval on A's, B's, and C's reference frames) is time dependent; in the case of B and C, each of the spaceships sees the three-dimensional distance between them as constantly increasing in time.

In the original proposal of the thread thought experiment ([Dew63], pp. 383–385), the breaking of the thread was attributed to the fact that, notwithstanding their simultaneous ignitions from A's perspective, B and C start their engines at different times from their own perspectives, and as a result the distance between them grows, or, as Bell puts it, B sees C drifting further and further behind (and, conversely, C sees B drifting further and further ahead) so that at a certain moment, when the velocities of the spaceships are high enough, the given piece of thread can no longer span that distance.[5]

One can look at things also from A's point of view. Once B and C are set in motion relative to A, the distance between the two spaceships will Lorentz contract, and the thread will break for the same reason mentioned above, namely, at each and every moment the velocity of the spaceships increases, and so the Lorentz contraction, as it were, is time dependent, and it increases at each and every moment, while the spaceships, with their increasing velocities, try to resist it and to maintain the distance between them constant (from A's perspective), until, at sufficiently

---

[4] Exactly when the thread breaks will depend on its mass, the mass of the spaceships, and their acceleration plan, see [Cor05].

[5] One should imagine two clocks that are situated at the two points where the thread is hooked to B and C. If these clocks are synchronized when B and C are at rest relative to A, then when B and C start moving with velocity v relative to A in the direction that is depicted in Figure 6.1, the clock at C will be ahead of the clock at B by lv/c^2 where l is the initial distance between B and C. The final separation between B and C will thus be equal to their initial distance plus the distance resulting from the difference in starting times.

great velocities, the thread will not hold the resistance of the spaceships, and will break.

Clearly, the acceleration of the spaceships involves forces, and these forces break the thread. But note that this fact does not contradict the view, germane to the principle approach of STR, in which both the relativity of simultaneity and the Lorentz contraction are considered kinematical, or geometrical, effects (a consequence of the structure of Minkowski spacetime and the different ways it can be "sliced" [Jan09]), and not dynamical effects. The breaking of the thread can still be explained kinematically, or geometrically, as a time-dependent projection of the four-dimensional interval between B and C at each and every frame. From the perspective of B or C, it is the fact that this time-dependent projection of distance (on the rest frame of each spaceship) increases due to the non-simultaneous (from that rest frame perspective) ignitions, so that at a certain moment the thread cannot span it. From the point of view of A it is the fact that the time-dependent projection of distance (on A's rest frame) is supposed to decrease while the spaceships maintain equal (and increasing) velocities, so at a certain moment the thread cannot withhold the stress. And so in both cases what explains the breaking of the thread is the structure of the Minkowski spacetime and the respective projections of the four-dimensional distance between B and C on each and every frame. A sees the thread contracting, pulling the spaceships together while they try to keep the distance between them fixed by increasing their resistance; B and C see the thread stretching, trying to keep the distance between the spaceships fixed, while they are constantly increasing it, hence pulling the thread apart.

How would the constructive approach explain the phenomenon of the breaking of the thread? On this view one regards the spaceships and the thread as material objects, obeying Lorentz covariant dynamical laws. Once set in motion from A's perspective, the material of the spaceships, and of the thread, will Lorentz contract: a sufficiently strong thread would pull the spaceships together and impose FitzGerald contraction on the combined system. But if the spaceships are too massive to be appreciably accelerated by the fragile thread, the thread has to break when the velocities become sufficiently great.

Recall that in the constructive approach the Lorentz–FitzGerald contraction is a dynamical process that is supposed, according to the proponents of that view, to explain the kinematics by introducing underlying Lorentz covariant forces. Situations such as Bell's thread, however, require careful attention since they can be easily misinterpreted: since the constructive approach identifies the Lorentz contraction as a dynamical process, it may look as if this process is the sole culprit in the breaking of the thread. But this is wrong. The Lorentz contraction alone does not account for the breaking of the thread, since if the spaceships B and C were not accelerating, but just drifting in constant velocities relative to A (and not accelerating from rest), nothing would happen to the thread that connects them, even though from A's perspective it would have Lorentz contracted. So while forces are clearly involved in the breaking of the thread, these forces originate in the acceleration of the spaceships; their explanatory role is thus limited to non-inertial scenarios and should not be confused with the role that the constructive approach assigns to the dynamical forces behind the kinematic Lorentz contraction (when no acceleration takes place).

Now some, apparently, believe that the non-dynamical and geometrical explanation given by the principle view of the breaking of the thread is mysterious:

What is required if the so-called spacetime interpretation is to win over this dynamical approach is that it offers a genuine explanation of universal Lorentz covariance. This is what is being disputed. Talk of Lorentz covariance 'reflecting the structure of spacetime posited by the theory' and of 'tracing the invariance to a common origin' needs to be fleshed out if we are to be given a genuine explanation here, something akin to the explanation of inertia in general relativity. Otherwise we simply have yet another analogue of Molière's dormative virtue. . . .

([Bro05b], p. 143)

From our perspective, of course, the direction of the explanation goes the other way around. It is the Lorentz covariance of the laws that underwrites the fact that the geometry of spacetime is Minkowskian.

([BP06], p. 84)

There are two ways to interpret these claims. On the first, strong reading, the complaint, in effect, is that Minkowski spacetime fails to supply a constructive (i.e., causal-dynamical) explanation of relativistic effects such as breaking of the thread, and for this reason the geometrical explanation is inferior to the dynamical explanation. On the second, weak, reading, Minkowski spacetime fails to supply a constructive (i.e., causal-dynamical) explanation of relativistic effects such as breaking of the thread, and for this reason it needs to be complemented with a dynamical explanation.

Now, on both readings, the strong and the weak, the first part of these claims is true. Recall that according to Einstein, on this view one is not engaged in postulating some hidden mechanism behind observed phenomena. Rather, the explanation of the breaking of the thread in Bell's thought experiment that this view subscribes to is very similar to the geometrical explanation that appears, for example, in [Put75] (pp. 295–296), where, if one wants to explain why a round peg cannot fit through a square hole, one points to geometric features of the peg and the board rather than solving the equations for the motion of all the atoms in the peg and the board. In the case of the thread, to repeat, this geometric feature is a consequence of the structure of Minkowski spacetime and the different ways it can be "sliced," i.e., the different ways the four-dimensional distance between B and C is projected on the three perspectives, or reference frames, A, B, and C.

Things turn out to be more complicated with respect to the second part of these claims. In Section 6.5.2 we shall examine the limits of a strong reading, according to which geometrical explanations are inferior to dynamical ones, and will recommend the weak reading, according to which geometrical explanations may be complemented with dynamical explanations, mere possibility proofs that a dynamical model is consistent with the geometrical one. On this weak reading, the dynamical and the geometrical explanations are on a par.[6]

Be that as it may, instead of delving further into a debate on the alleged superiority of one explanatory view over the other, where one man's horse is the other man's cart [BP06, BJ03], we shall set this matter aside and turn to examining Einstein's views on "length," as they emerge from his correspondence with Swann.

6.3 The Swann–Einstein correspondence

In this section I shall discuss a historical anecdote, mentioned briefly in [Bro05b] (pp. 119–120), that concerns a short correspondence Einstein had in early 1942 with the physicist W. F. G. Swann concerning the constructive approach to STR. I will use this correspondence to draw some interesting conclusions on the role of fundamental length in dynamical theories of spacetime.

6.3.1 Background

W. F. G. Swann (1884–1962) was born in England, and was educated at Brighton Technical College, the Royal College of Science, University College, Kings College and the City and Guilds of London Institute. Swann came to the USA in 1913 as head of the Physical Division of the Department of Terrestrial Magnetism at the Carnegie Institute in Washington. Later he was Professor of Physics at the University of Minnesota, the University of Chicago, and Yale, where he became Director of the Sloane Laboratory. He was appointed the Director of the Bartol Research Foundation in 1927.

A man of many talents, W. F. G. Swann was an accomplished cellist, the founder of the Swarthmore Symphony Orchestra, a former assistant conductor of the Main Line Orchestra, and former director of the Philadelphia Academy of Music. By the time of his appointment in Bartol, Swann had already distinguished himself as an excellent teacher, an outstanding researcher, and an emerging leader of the

---

[6] This view of constructive explanations as consistency proofs was prevalent among nineteenth century mathematical physicists such as Maxwell and Poincaré, and is even advocated by W. F. G. Swann ([Swa40], p. 276), another proponent of the constructive approach to whom the following sections are devoted. Swann, in his inimitable style, compares the relation between constructive explanations and their respective phenomenological principles to the logical justification that "our old grandmothers' remedies to our aliments as children" receive in terms of the fundamentals of bacteriology, physiology and chemistry.

scientific community. Although Swann is perhaps best known for his experimental and theoretical efforts in the area of cosmic ray physics, his research interests touched on many other disciplines such as condensed matter physics, relativity, and charged particle acceleration. In his capacity as a professor he is perhaps best known as the advisor of E. O. Lawrence who subsequently was awarded the Nobel Prize for developing the cyclotron. Lawrence followed Swann from Minnesota to Chicago, and then to Yale where he received his PhD.

Altogether Swann had over 250 publications including a popular science book *The Architecture of the Universe*. In 1967 the International Astronomical Union honored Swann when it gave his name to a crater on the lunar surface at 52 north latitude and 112 east longitude.

Swann advocates his constructive view on STR as early as 1912 in two papers with lengthy titles he published in the *Philosophical Magazine* [Swa12a, Swa12b]. He returned to this view years later in a series of papers published during the 1930s and 1940s in *Reviews of Modern Physics* (and in which he cites no one else but himself). It is this view which he also repeats in his correspondence with Einstein in 1942, and in his letter to Eddington that preceded it.

Swann's constructive approach to STR

Here is Swann in a representative quote:

If we strip from the theory [STR – AH] all the concepts incidental to its historical development; the fundamental outstanding dogma which remains is that the laws of nature – the differential equations – shall remain invariant under the transformation (1)–(4) [the Lorentz transformations – AH]. *Once we have written down some proposed laws*, the test of this conclusion is a matter of pen and paper, and not of experiment. . . . It is true that the principle of the restricted relativity owed its formulation to a belief that the coordinates associated with the various systems corresponded to the actual measures; but, once formulated, the working content of the theory, involving as it does the mathematical invariance of the laws, is independent of this hypothesis.

([Swa30], pp. 261–263)

Up to 1941 Swann's idea that the fundamental tenet of STR is the Lorentz covariance of the laws of physics was exemplified solely by electrodynamics. In 1941 he augmented his view with the claim that "relativity itself would provide no explanation of the [Lorentz–FitzGerald – AH] contraction were the story not capable of amplification by additional arguments based fundamentally upon the existence in nature of some such theory as the quantum theory" ([Swa41c], p. 197). In a footnote to this statement Swann mentions that already in his 1912 papers he had made the case for this claim when showing that the mere invariance of the electromagnetic equations, with the electrons regarded as singularities, was insufficient to explain the Lorentz contraction ([Swa12b], pp. 93–94). A relativistically invariant force equation was necessary in addition to the invariance of the field equation, but as time progressed it became more and more evident that "no obvious force equation following the lines of classical electrodynamics could be expected to provide the story of atomic and intermolecular forces in such a manner as to determine, ultimately, the form and equilibrium of a material body" ([Swa41c], p. 197, fn. 4).

Swann's intuition for the quantum nature of the cohesive forces in matter is mentioned already in [Swa41b], and the motivation for it already appears in 1912, but this intuition is spelled out in full for the first time only in [Swa41c], where he discusses the physical changes that a material rod will suffer when set in motion (I quote Swann extensively here since his correspondence with Einstein relies heavily on this passage):

When I start the rod in motion, all sorts of acoustical vibrations are set up. Of course, these will die down in time, but while I might be, perhaps unjustifiably, content if they should die down so as to leave the rod at its original length as measured in *S* [when the rod is at rest – AH], I am at a loss to know how the rod decides that it must settle down to a new length determined by the FitzGerald–Lorentz contraction. . . . It seems that quantum theory, if relativistically invariant in form, possesses the power to give the necessary answer. Consider the rod before the motion was imparted to it. What determines its form and stability? According to the quantum theory, these are determined by its being in a "ground state". Now the discussion given above . . . tells us that if the equations are invariant and we have in *S*, one solution for, let us say, the ψ function, satisfying the usual conditions of continuity, etc., then associated with this solution we have an infinite number of other solutions obtainable from it by a Lorentz transformation, and *these are all possible states in the system S*. . . . The ground state for our rod moving in a velocity *v* is the state obtainable by a Lorentzian transformation from the ground state of the rod before the motion was imparted.

([Swa41c], p. 201)

Here Swann has forcefully demonstrated the power of the constructive approach to STR ([Bro05b], p. 121). First, the relativity principle is the consequence of the Lorentz covariance of the quantum dynamics, rather than the other way round. Second, the universality of the behavior of rods and clocks emerges as a consequence of the dynamical argument, as long as matter of any constitution is assumed in principle to obey quantum theory.[7]

Detecting an opportunity to advocate his view to a larger readership, Swann repeated the above words almost verbatim in a letter he sent to *Nature* on September 23, 1941, which was also the pretext for his correspondence with Einstein on the subject.

---

[7] Several years later in a pedagogical paper published in the *American Journal of Physics*, Swann emphasizes this point when he explicitly constructs an "electron-dial" clock in his discussion on the twins paradox ([Swa60], pp. 59–61).

Swann's letter to Nature

In the second half of 1941 *Nature* hosted an exchange between the mathematical physicist J. H. Jeans and Sir Arthur Eddington. The former had just re-read Eddington's (1939) *The Philosophy of Physical Science* and was struck by Eddington's views on the *a priori* character of the fundamental laws of nature. Jeans reports that he had read Eddington "with great admiration, but also with grave doubts as to whether his philosophical position is not wholly unsound" ([Jea41], p. 140). The focus of the heated exchange that ensued between the two distinguished physicists in the pages of *Nature* (and which attracted the attention of others, e.g., Herbert Dingle) was the status of the light postulate of STR and the null result of the Michelson–Morley experiment. It is here where Swann enters the stage, and in a letter to *Nature*, written in September but published only in December that year ([Swa41a], p. 692), he spells out his own view on the role of the null result of the Michelson–Morley experiment in the foundations of STR.

Swann's view on this role sounds rather odd: he claims that there would be meaning to STR even if the Lorentz–FitzGerald contraction were not to hold, and the result of the Michelson–Morley experiment were non–null. What Swann meant here is that it is a purely mathematical fact that Maxwell's equations are Lorentz covariant. This fact, however, does not imply in itself that the transformed variables appearing in the Lorentz transformations refer to physical quantities actually measured by an observer moving at the appropriate velocity v relative to the original frame. After all, STR, on Swann's view, while securing the Lorentz covariance of the equations, does not tell us how rods and clocks behave in motion.[8]

Here Swann may have overstated his view ([Bro05b], p. 120): after all, on Swann's view, the Lorentz contraction is a result of the Lorentz covariance of the quantum theory of matter. But if the covariance of all the dynamical laws of nature is what follows from STR, as Swann claims, then Lorentz contraction must hold, if STR is to hold any physical meaning.

Replying to Swann, Eddington traces the solution to Swann's version of Brown's "mystery of mysteries" (i.e., "how a rod decides its extension when it is given a different motion" ([Edd41], p. 692)) to the law of gravitation. The connection to quantum theory, according to Eddington, is that it supplies us with a common standard for a measure of length, because "it is only in quantum theory that a method has been developed of describing material structure by pure numbers . . . Thus

---

[8] Swann goes on and restates his idea that if a co-moving observer adopts coordinates that match the moving rod (and clock) associated with a solution to the dynamical equations generated by the Lorentz symmetry group from the solution of these equations that describe a rod at rest, then the quantum description of the moving rod (and clock) by the co-moving observer is exactly the same as the description of the stationary rod by an observer associated with its rest frame.

appeal must be made to quantum theory for the definition of the interval ds, which is the starting point of relativity theory" ([Edd41], p. 693).

6.3.2 Swann's letter to Einstein

Swann, it appears, was not impressed with Eddington's reply, and a month after his letter to *Nature* was published, he wrote to Einstein (Swann's documented relation with Einstein had started a year earlier, when he tried to get an academic position for an acquaintance of Einstein, Felix Ehrenhaft).

In addition to mentioning his *Nature* letter, Swann enclosed one of his earlier *Reviews of Modern Physics* articles on the constructive view of STR from 1941, and asked for Einstein's opinion on this view. As can be seen from the page numbers to which Swann refers in his reaction to Einstein's reply (see below), the enclosed paper was [Swa41c]:[9]

I thought you might be interested in the enclosed in view of its relation of quantum theory to relativity. I gave a digest of it in *Nature*, December 6, and there is a comment by Eddington in the same issue. I find it very difficult to ascertain whether he agrees or disagrees. It seems that he agrees but does not like to.

As I see it, he wishes to deny the credit of the solution of the paradox to the quantum theory and give it to something more deep–seated, of which the quantum theory is an outcome. With this, of course, everyone must agree in principle, but the situation seems to me a little like one in which someone should give credit to a person A for a solution of a certain paradox, while somebody else claimed that it was A's ancestors to whom the credit should be given because it was they who were responsible for the existence of A.

Do not feel obligated to reply to this letter unless there is something you really wish to say about it, as I am only sending you the paper for your general interest.

(Swann to Einstein, January 3, 1942, quoted in [Hag08], p. 542)

Einstein's response arrived three weeks later and, as we shall see, contained some very interesting, albeit cryptic, remarks on the constructive view that Swann was advocating.

6.3.3 Einstein's reply

The German version

Einstein's response from January 24, 1942 to Swann's letter and to his *Reviews of Modern Physics* article (and maybe even to Swann's letter to *Nature*) was written in German and reads as follows:

Ich habe erst jetzt die mir freundlich übersandte Arbeit ansehen können. Mir schint deren Inhalt war nicht unrichtig, doch in gewissem Sinne irreführend.

---

[9] All quotes appearing here are verbatim, and are reprinted from the original letters that can be found in the W. F. G. Swann Archive at the American Philosophical Society, Philadelphia, PA. I thank Charles Greifenstein from the APS Library for his kind help in accessing the relevant letters.

In der speziellen Relativitäts-Theorie werden (idealisierte, aber doch im Prinzip als real-isierbar aufgefasste) Masstäbe und Uhren als selbständige physikalische Objekte behandelt, die – als verknüpft mit den Koordinaten der Theorie – in die Aussagen der Theorie mit eingehen. Dabei ist zunächst über die strukturellen Naturgesetze nichts weiter ausgesagt, als dass sie mit Bezug auf so definierte Koordinatensysteme Lorentz invariant sein sollen.

Es ist zunächst bewusst darauf verzichtet, Masstäbe und Uhren unter Zugrundelegung von Struktur-Gesetzen als Lösungen zu behandeln. Dies ist darum wohlbegründet, weil die (prinzipielle) Existenz solcher als Masstäbe für Koordinaten dienlicher Objekte vom Standpunkt unserer Erfahrungen besser begründet erscheint als irgendwelche besonderen Strukturgesetze, z.B. die Maxwell'schen Gleichungen.

Fasst man die Sache so auf, so hat der Michelsonversuch sehr wohl mit der speziellen Relativitäts-Theorie zu tun, sobald man das Prinzip von der Konstanz der Lichtgeschwindigkeit oder (darüber hinaus) die Maxwell'schen Gleichungen hinzunimmt.

Wenn man aber Masstäbe und Uhren <u>nicht</u> als selbständige Objekte in die Theorie einführen will, so muss man eine strukturelle Theorie haben, in welcher eine Länge fundamental eingeht, die dann zur Existenz von Lösungen führt, in denen jene Länge bestimmend eingeht, sodass es nicht mehr eine kontinuierliche Folge ähnlicher Lösungen gibt. Dies ist zwar bei der heutigen Quantentheorie der Fall, hat aber nichts mit deren charakteristischen Zügen zu tun. Jede Theorie, welche eine universelle Länge in ihrem Fundament hat und auf Grund dieses Umstandes qualitativ ausgezeichnete Lösungen von bestimmter Ausdehnung liefert, würde inbezug auf die hier ins Auge gefasste Frage dasselbe leisten.

(Einstein to Swann, January 24, 1942 (Einstein Archives 20–624). Published
(in German with an English translation) in [Hag08], pp. 543–544)

The English translation

Translated into English, Einstein's letter reads as follows:[10]

Only now have I been able to look at the work that you so kindly sent to me. It seems to me that its content was not incorrect, but still in a certain sense misleading.

In special theory of relativity measuring rods and clocks (idealized, but in principle conceived as realizable) are treated as independent physical objects, which, linked as they are to the coordinates of the theory, will enter into the propositions of the theory. At first there is nothing stated about the structural laws of nature other than the fact that they should be Lorentz invariant with reference to coordinate systems so defined.

Measuring rods and clocks are consciously not treated as solutions under the basis of structural laws [this sentence is a little ambiguous in the original]. This is well justified because from the point of view of our experiences, the (in principle) existence of those objects that can serve as measures for coordinates appears better justified than any particular structural laws, e.g. Maxwell's equations.

If one looks at the issue this way, the Michelson experiment does indeed have something to do with special theory of relativity, as soon as one adds the principle of the constancy of the velocity of light or (furthermore) Maxwell's equations.

But if one does NOT [underlined in the original] introduce rods and clocks as independent objects into the theory, then one has to have a structural theory in which a length is

---

[10] I thank Jutta Shickore and Sandy Gliboff for their help in translating Einstein's letter into English.

fundamental, which then leads to the existence of solutions in which that length plays a determinant [constitutive] role, so that a continuous sequence of similar solutions no longer exists. This is the case in today's quantum theory but has nothing to do with its characteristic features. Any theory that has a universal length in its foundation, and because of this produces qualitatively distinct solutions of a certain extension, would do the same with regard to the question under examination here.

Before we go on to decipher this response and to analyze its implications for some of Brown's claims, let us end this anecdote with Swann's reaction to Einstein's reply.

6.3.4 Swann's reaction

Swann seems to have asked for a translation of Einstein's letter, and this translation, which apparently was done by an amateur,[11] is enclosed in Swann's archives. One can only lament on this infelicity since Einstein's original German version of the letter is, to say the least, non-transparent and, as we shall see, an unqualified translation could easily (and actually did) lead to misunderstanding of Einstein's entire point.

Swann's reaction to Einstein's letter came a little more than a fortnight later, and it marks the end of his documented discussions with Einstein on the constructive approach to STR.[12] In his second letter Swann admits that the point he was making "did not depend specifically upon all of those features of the quantum theory which are of interest in the atomic structure," and he agrees with Einstein on the necessity of the constructive theory defining a length. Nevertheless, Swann claims that there are two additional reasons why one should look at quantum theory.

First, the theory displays a unique ability to supply a (relativistically invariant) measure of length (here Swann gives Bohr's hydrogen atom model as an example where the fundamental length would have been the radius of the first electronic orbit). Second, the theory allows us, using this measure of length, to determine what will be the ground state of the rod:

It is this power to fix a length which is a special aspect of the determination of structural form and that is why, in speaking of the rod, I give the quantum theory, on page 201 [of Swann's paper in the *Reviews of Modern Physics* from 1941 [Swa41c] – AH], the credit of determining that it shall be in a *ground state* [underlined by Swann – AH]. It seems to me

---

[11] The translation Swann got looks awkward to a native German speaker, and shows lack of knowledge of the context and subject matter.

[12] It appears that Swann maintained contact with Einstein and even visited him in 1950 in Princeton, to discuss a public talk he (Swann) was preparing at the Franklin Institute. No record of this meeting exists, apart from a letter Swann sent to Einstein after the meeting took place, to which he attached his planned talk, in order to thank Einstein and to allow him to veto any part of the talk that regarded Swann's reconstruction of Einstein's work on the theories of relativity.

that it is this act of the quantum theory which is significant and which is another aspect of its power to fix a length.

(Swann to Einstein, February 8, 1942, quoted in [Hag08], p. 545)

Swann ends his letter with an apologetic tone:

Please do not feel that it is necessary to reply to this letter. Of course, it is always a pleasure to hear from you on these matters, but I do not think there is any very great divergence of view point in this instance.

(Swann to Einstein, 8 February, 1942, quoted in [Hag08], p. 545)

from which it is clear that, ironically, Swann had interpreted Einstein's response as sympathetic to his (constructive) explanation of the Lorentz–FitzGerald contraction with a Lorentz covariant quantum theory of matter!

6.4 Reading Einstein

Admittedly, Einstein's letter to Swann is not one of his famous letters; nor is it one of his most transparent. In effect, while mentioned only twice in the vast literature on the foundations and history of STR, I believe that this letter was misunderstood on both occasions. In this section we shall reveal these misunderstandings, and then set the record straight.

6.4.1 Lost in translation

De-contextualization

In his recent *Einstein from 'B' to 'Z'* [Sta02], John Stachel devotes a chapter to "Einstein and the Quantum." Einstein's letter to Swann appears here for the first time in a rather misplaced context. Stachel brings the letter as evidence for the claim that Einstein (in his attempts to explore the possibility that a field theory based upon a continuous manifold, the principle of general covariance and partial differential equations, could provide an explanation for quantum phenomena) sometimes indicated that a fundamental length might be needed to explain the existence of stable structure in a field theory. He then (mis)quotes only the last paragraph of the letter ([Sta02], p. 394).

But clearly, by now the reader can appreciate that Einstein's letter to Swann, while it may be classified as pertaining to Einstein's views on the relations between relativity theories and quantum mechanics, is not in any way meant to indicate what Stachel asserts it does.

While the omission of the letter's first part (along with its context and attitude towards the constructive approach) is understandable from an editorial perspective, it appears to have led others to claim that Einstein – while skeptical about the role of quantum theory – nevertheless agreed with Swann that a dynamical explanation of the Lorentz–FitzGerald contraction was needed.

An alternative interpretation

Chapter 7 of [Bro05b], devoted as it is to the dissenting, unconventional, voices on relativity, is a wonderful starting point for any future research on the constructive approach to STR. Brown's inclusion of the Einstein–Swann correspondence in this chapter, however, is somewhat misleading. That Einstein expressed misgivings about his choice of the principle view is, of course, supported by textual evidence. I strongly doubt, however, that one can count Einstein's letter to Swann as such evidence (see below), but even if one chooses to do so, there exists another, more interesting problem. Reading Brown one gets the impression that Einstein was attracted to the constructive view of STR in general, but refrained from embracing Swann's version of it because of his suspicion of quantum theory.[13] Yet the evidence for this narrative is nothing else but the paragraph (mis)quoted by Stachel:

It is known that Swann corresponded with Einstein on the foundations of SR; in Stachel (1986) p. 378 [this part of the letter appears also in [Sta02], p. 394 – AH], part of a 1942 letter from Einstein to Swann is cited in which he discusses the possibility of a constructive formulation of the theory wherein rods and clocks are not introduced as 'independent objects'. Einstein argues in this letter that any such theory must, like the quantum theory, contain an absolute scale of length. It would be interesting to know more about this correspondence.

([Bro05b], p. 120, fn. 19)

6.4.2 A more accurate reconstruction

Reading Einstein step by step

Let us look more carefully at Einstein's response to Swann. The first thing to note is that from the very beginning, Einstein, in his famous style, is criticizing Swann without saying so explicitly when he writes in the opening sentence:

... It seems to me that the content [of Swann's work – AH] was *not incorrect but still in a certain way misleading* [my italics – AH].

One cannot refrain from making an analogy between Einstein's attitude to Swann's work and his reaction to Reichenbach's *The Philosophy of Space and Time* (1928), where Einstein, when asked whether he considers true what Reichenbach had asserted, mischievously answered "I can only answer with Pilate's famous question: 'what is truth?'" ([Sch49b], p. 676).

---

[13] Brown's view of this suspicion is also documented in [Bro05b] (pp. 187–190).

The second paragraph presents the principle view of STR, in which measuring rods and clocks are treated as idealized (but in principle realizable) primitives, i.e., independent, unanalyzed physical objects. On this view nothing is said about the dynamical laws of nature other than the fact that they should be Lorentz invariant with reference to coordinate systems defined by the primitive rods and clocks. Note that here Einstein restates the epistemological priority, under the principle view, of the geometrical symmetries of spacetime over the symmetries of the dynamical laws. To paraphrase Brown again, it is quite clear here which is the cart and which are the horses.

The next paragraph justifies this principle view: measuring rods and clocks are "consciously" (i.e., on purpose) not treated as solutions to the dynamical laws (i.e., solutions to the equations of motion) – as Swann would have it – and this view is well justified since on the basis of our experience (e.g., the null result of the Michelson–Morley experiment – see below – or the empirical content of the relativity principle), the *in principle* existence of these primitives as measures of coordinates (i.e., as representing the geometry of spacetime) seems to be better justified than any given dynamical theory of matter that one may come up with. This last sentence reminds us of the agnosticism of the principle view with respect to the actual dynamical model for relativistic phenomena.

However, in Swann's amateurish translation the first sentence of this paragraph was translated as saying "We at first consciously desist from treating measures and clocks as solutions by taking structural laws as basis." This was apparently interpreted to imply that instead of taking rods and clocks as primitives, as the former paragraph suggests, one should take the dynamical laws as a basis. Clearly this is not what Einstein meant here, as can be seen from the German version and from the place of this paragraph in the letter (Einstein turns to the constructive view only two paragraphs later).

Next comes another blow to Swann's view: if one looks at the issue this way (i.e., given the primitive character of the geometrical notions and the existence of rods and clocks), then the Michelson–Morley experiment does have something to do with the special theory of relativity, as soon as one adds the principle of the constancy of the velocity of light or (furthermore) Maxwell's equations.

It is only in the last paragraph where Einstein finally reverts to the constructive view. Unfortunately, this is the most obscure paragraph in the letter, and any minor changes in its translation can result in totally different interpretations.

The first sentence of this paragraph is a conditional sentence, with a short antecedent and a long and winding consequent. The short antecedent is, in paraphrase, preparing the stage for what, in Einstein's view, happens when one does not wish to take measuring rods and clocks as primitives, i.e., when one insists, contra all that was said before in the letter, in following the constructive approach to STR. The long and winding consequent, again in paraphrase, is a logical chain that follows from the constructive view. First, the dynamical theory that one utilizes in order to construct the kinematical effects (and to explain the geometry of spacetime) must establish a primitive measure of length. Second, this (here "this" is ambiguous – either this primitive measure of length, or the fact that the dynamical theory introduces a primitive measure of length) leads to the existence of solutions (of the dynamical equations) in which length plays a constitutive role. Third, since this primitive notion of length appears in the solutions to the equations of motion, a continuous sequence of similar solutions no longer exists (note that the negation "no longer exists" can be interpreted equally as quantifying over "continuous," or over "similar," or over both).

The second sentence says that the introduction of a primitive notion of length is the case with today's quantum theory but has nothing to do with its characteristic features. The final sentence then says that any theory which has a universal length in its foundations and because of this circumstance yields qualitatively distinguished or distinct solutions of a certain extent would do the same with respect to the question discussed here.

Taking stock

I shall divide my own reading of Einstein's response to Swann into four distinct parts (alternative readings are presented below in Section 6.5.4).

- The first is the opening sentence that sets the tone: Swann's view on the role of the Michelson–Morley experiment, and on the role quantum theory plays in dynamically explaining the Lorentz–FitzGerald contraction is, according to Einstein, misleading.
- The second is the description of the standard, principle, view of STR, where rods and clocks are taken as primitives, and the explanation of the Lorentz contraction is a geometrical one.
- The third is the warning that if one rejects the principle view and adopts the constructive view instead, then this move entails several non-trivial consequences: first, one must still designate dynamical entities in the theory as geometrical entities, and second, continuous symmetry of solutions is lost.
- The final part is the observation that *any* attempt to dynamically explain spacetime geometry (irrespective of quantum theory) will have the same consequences.

My first conclusion from this reconstruction is that, contrary to what Swann wanted to believe, in his letter Einstein was expressing little sympathy with Swann's view. This may be a warning against the broader claim Brown is making throughout his project regarding Einstein's views on the principle–constructive distinction. My second conclusion is more philosophical. Einstein's letter to Swann gives us a rare opportunity to discuss what Einstein saw as the limits of the dynamical approach to spacetime physics.

6.5　Einstein and the constructive approach to STR

6.5.1　*The physics of despair*

A recurrent theme in Brown's project is the claim that Einstein expressed, throughout the years from 1907 and at least until 1949, a certain "unease" with respect to his choice of the principle view of STR:[14]

> When Einstein formulated his 1905 treatment of relativistic kinematics, the template in his mind was thermodynamics. This was because a more desirable 'constructive' account of the behavior of moving rods and clocks, based on the detailed physics governing their microscopic constitution, was unavailable. The price to be paid was appreciated by Einstein and a handful of others since 1905.
>
> *([Bro05a], p. S85)*

Einstein's choice of the principle view, says Brown ([Bro05a], pp. S87–S89), was a choice of despair. Since several months before the publication of "The electrodynamics of moving bodies" Einstein had written another revolutionary paper claiming that electromagnetic radiation has a granular structure, the assumption, prevalent among his contemporaries, that Maxwellian electrodynamics is strictly true, had lost his trust. If Maxwell's equations were thought by Einstein as incompatible with the existence of the photon, then there was no sense in trying to write down a constructive, dynamical, theory for relativistic kinematical effects on the basis of classical electrodynamics as the latter could not be regarded as the complete theory of the constitution of matter. Einstein would have preferred a constructive account of the relativistic effects of length contraction and time dilation, but in 1905 the elements of an account of this sort were unavailable.

Now let us ignore Einstein's documented ambivalence with respect to the constructive approach, and suppose for the sake of argument that the question regarding Einstein's opinion is not only historically important but also unequivocally decidable. Let us even grant Brown's narrative above. Yet a puzzle still remains. If Einstein was so in favor of the constructive approach to STR and saw the principle view of 1905 as a choice of despair, then why, years later in his letter to Swann,

---

[14] Einstein's misgivings are voiced in Einstein's letter to Ehrenfest [Ein89a], his letter to Sommerfeld [Ein89b], his letter to the London *Times* [Ein54a], the famous paper on *Geometry and Experience* [Ein21], and several quotes from Einstein's reflections in his *Autobiographical Notes* [Sch49b]. In some of these quotes Einstein mentions that the example he had in his mind when retreating to the principle view of STR was thermodynamics, where, although no assumptions on the constitution of matter are made, a few phenomenological and restrictive principles suffice to predict the behavior of bulk matter.

after a relativistic quantum theory of matter had become available, was he still reluctant to acknowledge its implications on his choice?

One can sidestep this question by suggesting ([Bro05b], p. 114) that Einstein's "long-standing distrust" and hostility towards quantum theory prevented him from recognizing the progress in the theory and its implication for the formulation of STR. However, based on other writings I shall quote below, and on a certain interpretation of the last paragraph of his letter to Swann, I shall argue here that Einstein was troubled not with quantum theory *per se*, but with what he regarded as two general features of any constructive – dynamical – approach to spacetime physics, namely, (1) the idea that this approach would nevertheless require geometrical concepts, and (2) the conjecture that any attempt to explain dynamically the geometrical structure of spacetime would entail departure from the physics of the continuum.

6.5.2 The primacy of geometry

On the standard view of STR, as we understand it,[15] fundamental geometrical notions such as length are taken to be primitive. By length we mean what Einstein famously called *segment* (see Section 3.3.2), i.e., the distance between two spatial points (or persisting marks) on a rigid body (or intervals of local time):[16]

We shall call that which is given by two marks made on a practically rigid body a tract/a segment.

([Ein21], p. 237)

In this approach rods (and clocks) in STR are taken to fix the reference of geometrical notions such as "length," or "segment." No attempt is made in STR to give a dynamical analysis of rods and clocks that would result in depriving these notions of their primitive character.

According to Einstein ([Sch49b], p. 59), this use of rods and clocks in the standard view of STR introduces "two kinds of physical things, i.e., (1) measuring rods and clocks, (2) all other things, e.g. the electro-magnetic field, the material point, etc." This, he says, "is inconsistent; strictly speaking measuring rods and

---

[15] This section is partially based on joint work with Meir Hemmo [HH13].

[16] The German original is

Wir wollen den Inbegriff zweier auf einem praktisch-starren Körper angebrachten Marken eine Strecke nennen.

([Ein21], p. 391)

The standard translation in *Ideas and Opinions* has the idiosyncratic "tract" for "Strecke," but, given the context of Einstein's letter to Swann and the metrical reading of the term we suggest here, we believe that *segment* is an appropriate translation.

clocks would have to be represented as solutions of the basic equations [of motion], not as it were as theoretically self-sufficient entities."

Constructive (dynamical) approaches to relativity propose removing this inconsistency (Einstein, in [Sch49b], p. 61, calls it "a sin") by viewing the dynamics as primitive and as preceding geometry, and attempt to derive the Minkowski spacetime structure from the dynamical behavior of rods and clocks without any further input of primitive geometrical concepts. In this way they explicitly challenge the standard view of the special theory. The main idea at the core of this approach is to demonstrate the precedence of the dynamics over geometry.

While dynamical approaches to spacetime give priority to the dynamics over the geometry, it is not always clear in what sense they do so (see Section 6.1). In what follows we shall (partially) answer this question.

One way of understanding dynamical approaches, is that the features of the geometry of spacetime, such as the symmetries thereof, are derived from features of the dynamics. As an example of an approach of this sort, consider Maxwell's equations, and the statement that there are reference frames in which these equations hold. It follows mathematically that the collection of frames in which the equations do hold are related by the Lorentz transformation. Thus the dynamical theory uniquely picks the *symmetries* of the Minkowski geometry, while representing formally a non-geometric reality.

Another, stronger sense of this idea that seems to be associated with dynamical approaches is that not only the symmetries of the geometry, but also the very concept of spacetime as a geometric structure, can be fully derived from the dynamics, without assuming geometric notions such as length or volume. We will here argue that this stronger sense of deriving the geometry from the dynamics is untenable. At best one can show that the dynamics can be given a geometric interpretation. However, in our view, this does not amount to a strict derivation, or reduction, but rather to a sort of a consistency proof. Note that the weaker version of the dynamical approach is exactly a consistency proof in this sense: in the case of Maxwell's equations above there is no derivation of spacetime concepts in the stronger sense since one must have a clear sense of geometric concepts beforehand in order to construct the geometric interpretation out of the dynamical laws. Consistency proofs are a far cry from "derivations," and so in this stronger sense it seems to us misleading to think about the geometry as "emerging" from the dynamics alone.

Characterizing dynamical approaches in this stronger sense as we do, we clarify what the disagreement between the standard view and the dynamical approaches is about. Rods and clocks are obviously material bodies, and therefore a dynamical analysis of rods and clocks, and in particular of how they behave in measurements of, say, distance or time, is certainly to be sought for. Moreover, we concede that a dynamical analysis of the behavior of rods and clocks may teach us fundamental facts about the properties (i.e., the symmetries and transformation group) of the geometrical structure of spacetime. But it is crucial to note that this is not the issue of departure between the standard view of STR and the dynamical approach in the stronger sense. Rather, the issue at stake is whether one can dispense altogether with primitive geometrical notions such as "length," "area," "volume," etc., in the dynamical attempts to derive the geometrical structure of spacetime. That is, the issue at stake is whether one can derive notions such as "segment" from the dynamics alone.[17]

In Section 3.3.2 I have already presented the argument for the untenability of the stronger sense of the dynamical approaches: any approach of this sort is deemed epistemically incoherent unless it designates a theoretical entity as a geometrical measure of length, as without a primitive geometrical notion, the theory is unable to make contact with experience, and is therefore unverifiable. What we suggest here is that in his correspondence with Swann, Einstein makes the same point.

We iterate that by a primitive geometrical interpretation, we mean primitive notions, such as "length," "area," and "volume," etc., as measured by rods and clocks; we do not mean the symmetries or transformation group of the geometrical structure. This, we believe, is the content captured by Einstein's notion of "segment." This notion does not imply anything about the properties and the symmetries or transformation group of the background geometry of spacetime; this question of the full geometric structure of spacetime is contingent and is a matter of empirical discovery. Accepting a primitive notion of "segment" is minimal; it must be presupposed no matter what the geometrical structure of spacetime turns out to be, and it only means that one is concerned with a geometrical structure, whatever its symmetries are.

If this point is correct, then "segments" are not the sorts of things that can be derived from dynamical considerations. They remain fundamental and primitive with or without a dynamical underpinning of rods and clocks, and they are required before a dynamical story can be tested empirically. The dynamical analysis of rods and clocks must include some theoretical notions, that, by construction (or stipulation), are taken to refer to the primitive geometric notion of "segment." Without this stipulation it seems impossible to connect the predictions of the theory with the sorts of things we can measure, i.e. with our experience. Consequently, according to this view, a successful dynamical analysis can at most be taken as demonstrating the consistency of the dynamics with our measurements of length

---

[17] Apart from Swann, for recent proponents of the dynamical approach in the context of STR see, e.g., [BP06, Bro05b, Hug09]. We are unsure to which nuance, the strong or the weak, each of them adheres, but it seems that the best way to understand these authors is along the lines of the weak sense. The purpose here is to reveal the limits of the stronger sense of the dynamical approach, as the literature is often ambiguous on this issue.

and time. It cannot be understood as conceptually preceding the concepts of length and time.[18]

This view, we suggest, may explain Einstein's point when he says (in the second paragraph of his letter to Swann above) that rods and clocks (which we interpret in terms of segments) are better justified than, for example, Maxwell's equations. In this sense, the standard view of STR cannot be replaced by any dynamical view *if* by "replacement" one means that the dynamics precedes (conceptually or in any other sense) the geometry. This is why we think that Einstein saw no genuine contradiction between the standard view (which he himself developed) and the aforementioned "sin." We therefore ought to think about Einstein's "sin" not as a mark of inconsistency but rather as a mark of incompleteness.

As we shall see in Chapter 7, current dynamical approaches, as a matter of fact, presuppose a primitive geometrical notion of length (or area, or volume), rather than derive it from the dynamics. All of them fix the reference of the notion of "segment" by picking out some theoretical magnitude that is said to designate the (primitive) measure of length. This move is interpretative and does not amount to a derivation of the primitive geometry from the dynamics. Rather, one has to have in advance a primitive notion of "segment" in order to make this move.

We shall exemplify this point here with Swann's proposal, but we believe it is much more general. The analysis below will demonstrate that the stronger reading of the dynamical approaches is simply an overkill: the attempts to construct geometry from dynamical considerations cannot deprive the primitive geometrical notions of their primacy. Einstein's aforementioned sin cannot be interpreted as implying more than a desire for completeness (or consistency between the dynamics and the results of measurements carried out by means of rods and clocks). In this sense, *there can be no purely dynamical derivation of geometry.* Let us call this statement "thesis *L*" (for "length").

Swann's small rod

Let us test thesis *L* and see whether it applies to Swann's proposal. Swann (above) suggests taking the quantum mechanical ground state of the rod as determining its rest length. But as Einstein points out (above, and Swann himself later agrees) this move will be pointless unless some quantum mechanical magnitude will stand for the primitive measure of length. Swann concedes, and suggests the radius of

---

[18] Note that this argument does not mean that reduction is impossible *simpliciter*; one can still dynamically derive the notion of temperature from, say, the kinetic theory of gases, by defining it as mean kinetic energy. The point is that, contrary to the attempt to derive geometry from dynamics, the predictions of the kinetic theory of gases can be empirically verified without any recourse to the notion of temperature, as, under the assumption that all measurements are position measurements, we actually measure "segments" when we measure temperature.

the first electronic orbit in Bohr's model of the hydrogen atom as designating this measure.

The crucial point to be made here is that in introducing the Bohr radius as the primitive measure of length, Swann already presupposes a primitive notion of "segment" as designating, or fixing, the length of a smaller rod defined by the Bohr radius. Note, also, that the Bohr radius is picked out as designating a measure of length explicitly in John Bell's Lorentzian approach to STR [Bel87].[19] In this sense, therefore, the dynamical approach to STR reduces back to the standard view, albeit, as it were, on a different (in this case, smaller) scale.

Here we should stress two points. First, when we say that Swann presupposes a primitive notion of "segment" by "fixing the length" of a smaller rod as given by the Bohr radius we do not mean that he merely fixes a scale for the theory. Rather we say that he fixes the reference of the notion of length. That is, Swann identifies a theoretical magnitude (i.e., the Bohr radius as given by the ground state of the relativistic quantum theory) with the reference of the geometrical notion of "segment." In other words, Swann presupposes that, for example, the Bohr radius is the segment, just as Einstein presupposes in the standard formulation of STR rods (and clocks) as primitive notions. It is for this reason that we say that the notion of "segment" precedes the dynamics.

Second, a proponent of the dynamical approach may say that the Bohr radius (for example) is not *a priori* geometrical and only gets its geometrical status in virtue of its role in dynamical solutions that serve as rods and clocks. However, this does not counter thesis *L*, since the crucial point is that it is only by stipulation that one associates certain solutions of the dynamical equations with rods and clocks. In other words, the dynamical equations of motion only get their geometrical status in virtue of the fact that their solutions may be interpreted as referring to rods and clocks, no matter at what scale these rods and clocks are being used.

Recall that we distinguished between the question of stipulating the notion of "segment" and the question of the properties (i.e., the symmetries and the transformation group) of the geometrical structure of spacetime. So far we have only addressed the former. As to the latter, the dynamical approach faces the danger of vicious circularity in the case of constructing STR from quantum field theory. The reason is this.

In Swann's proposal, for example, the primitive notion of length, designated by the Bohr radius, is by construction relativistic invariant, since relativistic quantum

---

[19] We therefore do not interpret Bell's famous paper as an attempt to derive the geometry from the dynamics, nor as replacing the standard view by a dynamical view of STR, but rather as an attempt to complete the standard approach with a dynamical account, consistent with our measurements with rods and clocks. Note that even according to Bell's "Lorentzian pedagogy," in the single frame we use to describe the dynamical behavior of a rod we still require a notion of "segment" (which, in this case, is frame dependent).

theory (which later became quantum field theory) presupposes from the outset, as a matter of empirical generalization, the Lorentz covariance of the equations of motion. But this means that in order to derive the relativistic effects from the dynamics of quantum field theory, one must impose Lorentz covariance as a global feature on all the non-gravitational interactions. The point here is that, first, Lorentz covariance is imposed rather than derived, and second, in this case it is imposed as a constraint on the theoretical magnitude that was picked out as designating the fundamental measure length. But, as we argued above, since with the said designation the fundamental measure of length is ultimately a geometrical notion, Lorentz invariance is an assumption about the geometry of spacetime; it is not an assumption about the dynamics *per se*.

We therefore agree with Mermin ([Mer05], p. 184) that the quantum field theoretic construction of STR is circular, since in this construction the assumption of Lorentz covariance carries all the work in the proofs of the relativistic effects. Pauli ([Pau21], p. 15) seems to acknowledge this constraint independently of quantum mechanics. In both cases the assumption of Lorentz covariance is generalized from empirical facts that were discovered using "segments" (primitive rods), and so according to thesis *L*, the circularity is even deeper, since this assumption is, on final account, a structural assumption about the geometry. This means that before a geometrical interpretation is put forward, one has to make a decision about the geometry, for example, about whether or not there are rigid bodies in the world. Of course, one's decision should be consistent with one's theoretical background and experimental support of the overall theory. We take this to reiterate Poincaré's main point about the geometry in his famous disk argument (see Section 3.3.1). We thus suggest reading Einstein's closing comments in the third paragraph quoted above in light of this analysis.

6.5.3 Departing from the continuum

There is no doubt that the principle–constructive dichotomy was a major theme in Einstein's view on STR, and Brown [Bro05b] is an excellent example of the fruitfulness of the attempts to reconstruct this view within its framework. But there exists a parallel, even more important, dichotomy with which one can describe Einstein's lifework, namely, continuum-based or discontinuum-based physics. As Holton [Hol72] and Stachel [Sta93] note, at first glance most of Einstein's work pertains to the physics of the continuum in both its physical and its philosophical aspects: his conclusion, based on STR, that direct-interaction theories are no longer tenable once one recognizes the existence of a maximum signal velocity ([Sch49b], p. 61), his forty years long search for a unified field theory of gravitation and electromagnetism, and the field theoretic account of gravitation given in the general theory of relativity (GTR). On the other hand, discontinuity and discreteness also played an important role in Einstein's work, for example, classical atomism as embodied in his work on Brownian motion, his light quanta hypothesis, and his contribution to the Bose–Einstein statistics.

Stachel ([Sta93], p. 276) remarks that the spacetime arena unifies the continuum and the discrete, since both (continuous) fields and (discontinuous) atoms find their home in the spacetime continuum. This leads him to discuss what Einstein saw as the alternative to continuum-based physics, an alternative which purports not to take spacetime as a primitive. Einstein referred to this physics as "purely algebraic":

… one does not have the right today to maintain that the foundation [of theoretical physics – AH] must consist in a *field theory* [emphasis in the original] in the sense of Maxwell. The other possibility, however, leads in my opinion to a renunciation of the spacetime continuum and to a purely algebraic physics. … For the present, however, instinct rebels against such a theory.

(Einstein to Paul Langevin, October 1935, in [Sta93], p. 285)

To be sure, it has been pointed out that the introduction of a space-time continuum may be considered as contrary to nature in view of the molecular structure of everything which happens on a small scale. It is maintained that perhaps the success of the Heisenberg method points to a purely algebraical method of description of nature, that is to the elimination of continuous functions from physics. Then, however, we must also give up, by principle, the space-time continuum. It is not unimaginable that human ingenuity will some day find methods which will make it possible to proceed along such a path. At the present time, however, such a program looks like an attempt to breathe in empty space.

([Ein54b], p. 319)

The alternative continuum–discontinuum seems to me to be a real alternative; i.e., there is no compromise. By discontinuum theory I understand one in which there are no differential quotients. In such a theory space and time cannot occur, but only numbers and number-fields and rules for the formation of such on the basis of algebraic rules with exclusion of limiting processes. Which way will prove itself, only success can teach us.

(Einstein to H. S. Joachim, August 1954, in [Sta93], p. 287)

In his essay "The Other Einstein" Stachel exposes Einstein's ambivalence towards the physics of the discontinuum, or the discrete, and it is against this background that I suggest further reading Einstein's letter to Swann. For apart from thesis L and the epistemological argument about the primacy of geometrical notions they must face, Einstein seems to suggest that constructive approaches of the sort Swann was promoting would eventually lead us away from continuum-based physics.

Exactly when Einstein started to consider the problem of formulating statements about the discrete, without calling to aid the continuum spacetime, is still unclear. Stachel [Sta93] cites Einstein's letter to Walter Dällenbach from November 1916

as the first direct evidence (see Section 5.2), but suggests ([Sta93], p. 281) that
Einstein may have started to think upon this question well before his work on GTR.

Note that on this view Einstein's reaction to Swann's letter is definitely not
meant to express hostility to quantum theory *per se* on the basis of his "long-
standing distrust," nor should it be interpreted as expounding the (false) claim that
quantum theory is inconsistent with the continuum.[20] Rather, I believe that here
Einstein is expressing his worries about the consequences of the attempt to construct
consistency proofs for the structure of spacetime from a more fundamental theory.
For discontinuum-based physics would be the kind of physics Swann would end up
with if he refused to take the structure of spacetime as fundamental and exchanged
the explanans for the explanandum, no matter what dynamical theory he used.

I realize, of course, that the temptation to use Einstein's hostility to quantum
theory as a "catch-all" narrative is strong: it is well known that Einstein was
accused of "rigid adherence to classical theory," and similarly to Brown, most
commentators interpret this rigidity as manifest in Einstein's view of quantum
theory as incomplete (Bohr in [Sch49b], p. 235), or in his reluctance to abandon
the notion of separability [How85], or even in his prejudice for causality ([Fin86],
pp. 100–103). But the correct interpretation of this rigidity was given by Einstein
himself:[21]

The opinion that continuous fields are to be viewed as the only acceptable basic concepts,
which must also [be assumed to] underlie the theory of material particles, soon won
out. Now this conception became, so to speak, "classical," but a proper, and in principle
complete, *theory* has not grown out of it. . . . Consequently there is, strictly speaking, today
no such thing as a classical field-theory; one can, therefore, also not rigidly adhere to
it. Nevertheless, field theory does exist as a program: "Continuous functions in the four-
dimensional [continuum] as basic concepts of the theory." Rigid adherence to this program
can rightfully be asserted of me.

([Sch49b], p. 675)

6.5.4 Alternative interpretations

Given the obscurity of the last paragraph in Einstein's letter to Swann, it can be
given different interpretations. I suggest two of them here and leave it to the reader
to decide between them and the one I have suggested above.

The first alternative sees this paragraph as referring to Einstein's own attempts to
arrive at a field-theoretic construct for the fundamental particle structure of matter.
The claim here is that it had been one aspect of Einstein's (and others') unification
program since the inception of GTR to find some generalized or modified set of

---

[20] If one assumes that quantum dynamics is enacted on a background spacetime continuum, then of course it is
consistent with the continuum.

[21] In this passage Einstein directly responds to the "friendly accusations" made by Pauli ([Sch49b], p. 158).

field equations that would allow the derivation of static, spherically symmetric, solutions that one could then interpret as matter particles [Sau07].

Specifically, one requirement of this program had always been to account, in this way, for the existence of an electron and a proton of quantized charge and with a given mass difference. Therefore, it was expected that the field solution of a particle would come out somehow quantized and not with a continuous parameter, that would label "similar" yet different solutions. Einstein's point, on this view, is that one would expect that somehow the electron charge and mass come from a definite specification of the field configuration that gives its representation, and that this specification naturally would provide a fundamental length. Under this interpretation, the program Einstein had in mind in his letter to Swann was the program of "explaining" matter (i.e., the existence of electrons and protons with their given charges and masses) from a fundamental underlying (continuum) field theory, encapsulated in unified field equations of gravitation and electromagnetism.

This view seems to agree with Stachel's (see Section 6.4.1). I remind the reader, however, that this view takes Einstein's remark to Swann and puts it entirely out of context. As this chapter shows, the background with respect to which the Einstein–Swann correspondence took place was not "the explanation of matter," but rather the explanation of the structure of spacetime with a specific dynamical theory of matter.

The second alternative reconstructs Einstein's reply to Swann as follows: in order to account for length contraction and time dilation, a constructive theory must not be conformally invariant, since otherwise no notion of scale would emerge. Quantum mechanics has a natural length scale (and is not conformally invariant), but quantum mechanics is not the last word. As long as some constructive theory comes along which is non-conformally invariant, it will do the job just as well. That is why Swann's insistence on the central role of quantum theory in his constructive account is "in a certain sense misleading."

This interpretation agrees with mine, that a constructive approach to STR must introduce a fundamental length, yet, and it is here where the disagreement lies, the word "fundamental" is interpreted differently (according to this reading it means "absolute"; according to mine it means "primitive"). Moreover, according to this view, Einstein's remark to Swann does not exemplify his worries about the consequences of this fundamental length (a reading, however, which naturally follows from the structure of the paragraph in its German original). Rather, it explains why Einstein was not that impressed with Swann's particular choice of quantum theory.

What was the true reason behind Einstein's remark to Swann, I am afraid, will remain a matter of contention.

6.6 Geometry and dynamics, again

Relying on textual evidence, in this chapter I have argued that while Einstein's 1905 choice of the principle view of STR may have signified a choice of despair, his reluctance to embrace the constructive approach in his later years can be interpreted as a "rigid adherence" to primitive geometrical notions and to continuum-based physics.

Setting these historical issues aside, and regardless of what Einstein meant in his letter to Swann, there exists an interesting philosophical question, namely, what are the consequences of explaining geometry with a dynamical theory of matter, or of pursuing a constructive approach in the context of spacetime theories?

I think it is safe to say that we can now identify at least one possible consequence – our thesis L – that says that any dynamical approach to spacetime geometry is epistemically incoherent unless it designates a theoretical magnitude as geometrical, so predictions of the theory could be verified in spacetime. This interpretation sees Einstein's reply to Swann as defending the primacy of primitive geometrical notions against attempts to derive them from any dynamical theory of matter – presumably more fundamental.

That thesis L follows from an operationalist philosophy was argued for in Section 3.3.2. In this chapter I have suggested that Einstein's correspondence with Swann could be read as lending support to this thesis. The remarkable fact is that, as we shall see below, the current landscape of quantum theories of gravity also vindicates this thesis: similarly to Swann in the context of STR, current contenders for a solution of the quantum gravity problem do, in effect, designate a primitive geometrical notion of length. Doing so, they also vindicate Einstein's second worry, namely, that the attempts to approximate spacetime geometry with a more fundamental dynamical theory would eventually lead us away from continuum-based physics.

7

Quantum gravity: current approaches

7.1 Outline

Classical general relativity (GTR) and the Standard Model are two spectacular theories whose empirical success is unmatched throughout the history of science. Using them, we are able to account for virtually anything we can measure (except, maybe, for dark matter). The theory behind the Standard Model, quantum field theory (QFT), provides a general framework for all theories describing particular interactions of (quantum) matter fields. GTR, on the other hand, summarizes what we know about gravity and spacetime at the large scale. And yet these two theories do not allow us to make predictions in all physical regimes. The only interaction which has not been fully accommodated under the QFT framework is the oldest known one, namely, the gravitational interaction. Specifically, if we want to compute the scattering amplitude of two point-particles interacting gravitationally, we can do so using (non-renormalizable) perturbative quantum general relativity as an effective theory, but predictability breaks down when the center of mass energy of the particles becomes of the order of the impact parameter in Planck units (see below). In other words, current established unification of QFT and GTR gives no prediction whatsoever of what happens to particles that scatter at that energy.

The problem of unifying QFT (which describes all non-gravitational interactions) with GTR (which describes, via Einstein's field equations, gravitational phenomena) is known as the quantum gravity problem. This unification, at least in its scientific facet, is supposed to enable predictions, in a logically consistent way, for situations in which *both* gravitational effects and particle physics QFT effects cannot be neglected. Its domain of applicability includes effects which are germane to the short-scale structure of physical space, to events in early cosmology, and to some aspects of black hole physics.

The scale at which the aforementioned quantum gravitational effects become dominant is denoted by the so called Planck units, based on the three fundamental constants G, c, and $\hbar$ that can be combined in a unique way to yield units of length, time, and mass:

$$l_P = \sqrt{\frac{\hbar G}{c^3}} \approx 1.62 \times 10^{-33} \text{ cm,} \tag{7.1}$$

$$t_P = \frac{l_P}{c} = \sqrt{\frac{\hbar G}{c^5}} \approx 5.40 \times 10^{-44} \text{ second,} \tag{7.2}$$

$$m_P = \frac{\hbar}{l_P c} = \sqrt{\frac{\hbar c}{G}} \approx 2.27 \times 10^{-5} \text{ g} \approx 1.22 \times 10^{19} \text{ GeV.} \tag{7.3}$$

Planck [Pla99] was among the first to note that

These quantities retain their natural meaning as long as the laws of gravitation, of light propagation in vacuum, and the two laws of the theory of heat remain valid; they must therefore, if measured in various ways by all kinds of intelligent beings, always turn out to be the same.

(Planck, translated in [Kie07], p. 5)

Almost two generations later, around 1955, John Wheeler, in discussions with his student Misner, coined the term "quantum foam," and designated the Planck length (and the Planck time) as the scale where quantum gravity effects may take place:

What emerged from my discussions with Misner was the realization that it is this "Planck length" that sets the scale of quantum fluctuations in spacetime. It is quite incredibly small. How small? Imagine a line of spheres, each with a diameter equal to the Planck length. The number of these spheres that would have to be lined up to span the diameter of one proton would be the same as the number of protons that would have to be lined up to reach across New Jersey. It takes 100,000 protons in a row to equal the size of one atom, and 1 million atoms in a row to go from one side to the other of the period at the end of this sentence. You don't need to be told that it would take a lot of these periods to stretch across New Jersey. Relative to the Planck length, even the minuscule entity we call an elementary particle is a vast piece of real estate. To change the analogy from distance to money, a penny relative to the U.S. annual budget is a million times larger than the Planck length relative to the size of a proton. Just as there is a Planck length, there is also a "Planck time". It is the time needed for light to move through one Planck length. If a clock ticked once in each Planck unit of time, its number of ticks in one second would be greater by many billion fold than the number of vibrations of the quartz crystal in your watch during the lifetime of the universe. The Planck time is simply too short to visualize – but not too short to think about! If interesting things happen within the dimension of a Planck length, they also happen within the duration of the Planck time.

([FW98], pp. 245–246)

Wheeler also believed that at that scale ordinary spacetime concepts would break down:

On our imaginary downward voyage to ever smaller domains, after reaching the size of a single proton, we would have to go twenty powers of 10 further to reach the Planck length. Only then would the glassy smooth spacetime of the atomic and particle worlds give way to the roiling chaos of weird spacetime geometries. The wormhole would be but one simple manifestation of the distortions that could occur. So great would be the fluctuations that there would literally be no left and right, no before and no after. Ordinary ideas of length would disappear. Ordinary ideas of time would evaporate. I can think of no better name than quantum foam for this state of affairs.

([FW98], pp. 246–247)

These ideas would inspire many physicists, and as we shall see below (Section 7.5), would culminate in some astonishing claims about "the disappearance of spacetime" at the Planck scale.

While the Planck mass m_P seems to be large for microscopic standards, one needs to recall that this mass (energy) needs to be concentrated in a region of linear dimension l_P for the quantum gravitational effects to become dominant. This means that such effects can be attained for elementary particles whose Compton wavelength is (up to a factor of 2) equal to the Schwarzschild radius,

$$\frac{\hbar}{m_P} \approx R_S \equiv \frac{2Gm_P}{c^2}, \tag{7.4}$$

i.e., when the spacetime curvature that the elementary particle inflicts is non-negligible.

As Wheeler succinctly indicated, the Planck length is very small, almost 20 orders of magnitude smaller than the radius of the electron. This seems to suggest that the relevant scale for quantum gravitational effects is well beyond the range of any foreseeable laboratory-based experiments, and that the only physical regime where effects of quantum gravity might be studied directly is in the immediate post big-bang era of the universe – not the easiest thing to probe experimentally [Ish95].

The lack of empirical evidence notwithstanding, theoretical frameworks still flourish. In this chapter we shall survey three current strategies, or approaches, to the quantum gravity problem, focusing in particular on the ways in which the notion of fundamental length is introduced (for two recent technical reviews see [Gar95, Hos13]). As we shall see, our thesis *L* about the primacy of geometry (Sections 3.3.2 and 6.5.2) is strongly vindicated: in all these strategies specific theoretical entities are chosen to designate primitive geometrical notions of measure (of distance, area, or volume).

The first strategy stems from particle physics, where one is tempted to see gravity simply as one more gauge interaction, and a natural solution to the quantum gravity problem would be String-theory-like: a theory whose core features are essentially described in terms of graviton-like exchange in a background classical (Minkowski) spacetime (see, e.g., [Smo01], pp. 179–193). Next, from the general-relativistic perspective it is natural to renounce any reference to a background spacetime [Smo05], and to describe spacetime in a way that takes into account the quantum mechanical and gravitational limitations on measurements. The Loop Quantum Gravity (LQG) approach [Rov04] and the causal set approach [BLMS87] originate from this perspective. Finally, the third possibility is a condensed matter perspective (see, e.g., the research programs of [Vol03] and [Lau03]) from which some of the familiar properties of spacetime are supposed to emerge.

7.2 String theory

String theory is an attempt to solve the quantum gravity problem, and to do so within a unified (quantum) framework of all interactions. I am not an expert on Strings. My aim in this section is not to introduce the reader to the mathematical intricacies and the manifold of physical aspects that almost five decades of hard work by hundreds of brilliant minds have generated (for general background see, e.g., [Pol98, Zwi08]; for String theory as a theory of quantum gravity see, e.g., [BT09, Muk11] and [Kie07], Ch. 9). Rather, I shall focus here on a very basic question, namely, what entity plays the role of the primitive notion of length in String theory, and how does this role relate to the history that was surveyed in earlier chapters.

7.2.1 The string length and its role

String theory treats gravity (and also the other gauge interactions) as mediated by massless spin two (gravitons) and spin one (gauge boson) particles, which correspond to closed and open strings, respectively ([Ven92], p. 140). String theory thus provides a unifying framework of all elementary particles and their interactions: it automatically includes gravity (in the form of a massless traceless symmetric second-rank tensor excitation of the closed string, identified with the graviton) in addition to gauge forces which arise from massless excitation of the open or closed string (depending on the perturbative formulation of the theory) [BT09].

Reducing as it does all interactions to excitations of closed and open strings, String theory proudly achieves a finite perturbative description of quantum gravity. The price, however, is additional structures: extra dimensions, compactification as for the Kaluza–Klein theory, infinite towers of massive fields, and supersymmetry,

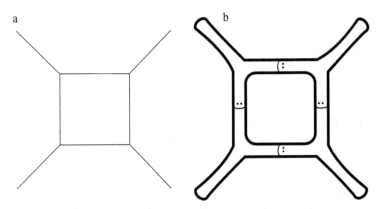

Figure 7.1 Removing singularities with strings.

to name a few. A lot of ink has been spilled on the methodological pros and cons of these additional structures (e.g., [Smo01, Woi06, Pol07]). Here we shall simply note that one of these extra features is singled out as presumably allowing String theory to remove the ultraviolet divergences that plague QFT, and also the singularities that haunt GTR: the extended nature of the string is supposed to serve as an effective cutoff for these divergences ([Wit96], p. 25; [Rov11]).

That the basic building blocks of nature are extended and not point-like, and that this extension is instrumental in removing ultraviolet divergences, are premises that can be found in most introductions to String theory. For example, in a semi-popular exposition, Ed Witten [Wit96] presents String theory as an elaboration on quantum theory of gravity that follows Feynman's rules (for an early application of Feynman's rules to the quantization of gravity see [Mis57]). Feynman's diagrams are then "smoothed out" by replacing the one-dimensional world-line of a point-particle with "world-tubes," two-dimensional orbits of strings. In Witten's words, and as can be seen in Figure 7.1 (taken from [Wit96], p. 27), "There is no longer an invariant notion of when and where interactions occur, so from the description above of the origin of the problems of field theory, we might optimistically hope to have finiteness. . ." ([Wit96], p. 25).

Other String theorists are even more explicit about the role of this new extended entity:

In String theory, the Planck constant λ_s [the string's length – AH] is also the short distance cut-off. This is so because, as we have seen, strings acquire, through quantization, a finite, minimal size $O(\lambda_s)$. The finite size induces in turn a cut-off (in the way of a form-factor) on large virtual momenta.

([Ven92], p. 141)

The fundamental length in String theory is thus the length of the string, or rather, the string tension $(2\pi\alpha')^{-1}$, with a new fundamental parameter $\alpha' = 10^{-32}$ cm^2 of dimension length squared:

$$l_s = \sqrt{2\alpha'\hbar}. \tag{7.5}$$

The role this characteristic length plays in removing the ultraviolet divergences is very similar to the role played by the fundamental length in the theories we have encountered in Chapter 4. Moreover, according to String theory, this length also delimits the resolution of our spatial measurements:

> There is a smallest circle in String theory; a circle of radius R is equivalent to a circle of radius α'/R. . . . Imagine that the universe as a whole is not infinite in spatial extent, but that one of the three space dimensions is wrapped in a circle, making a periodic variable with period $2\pi R$. Then there is a smallest possible value of R. When R is large, things will look normal, but if one tries to shrink things down until the period is less than $2\pi\sqrt{\alpha'}$, space will re-expand in another "direction" peculiar to String theory, and one will not really succeed in creating circles with radius less than $\sqrt{\alpha'}$.
>
> *([Wit96], pp. 28–29)*

Here "direction" means the direction of contraction/expansion of the resolution, i.e., the decrease/increase in the scale of the observation. This feature of String theory is also known as *T-duality*: the physics at radius R is the same as the physics at radius α'/R (see, e.g., [Pol98], Ch. 8).

This limit suggests, according to String theorists, a modification of Heisenberg's uncertainty relation to:

$$\Delta x \geq \frac{\hbar}{\Delta p} + \alpha' \frac{\Delta p}{\hbar}, \tag{7.6}$$

where the first term is the familiar, quantum theoretic one, and the second is the contribution due to the string uncertainty. This modified uncertainty relation implies an absolute minimum uncertainty in length measurement of the order $\sqrt{\alpha'} \approx 10^{-32}$ cm in any experiment.

It is interesting that in this context there already exist two related, model independent, results on the limitation of precision of geometrical measurement. Both these results stem from the relation between the area of a black hole and its mass. We have already encountered the first, which goes back to Alden Mead (see [Mea64] and Section 5.5.2), where the uncertainty is due to the formation of black holes if one uses probes of too high an energy, which limits the possible precision:

$$\Delta x \gtrsim \sqrt{G}, \tag{7.7}$$

where G is Newton's constant. If one requires in addition that the particle's momentum uncertainty Δp should be of the order of the probe's momentum, and that the familiar Heisenberg uncertainty and the gravitational uncertainties add linearly, one arrives [AS99] at:

$$\Delta x \gtrsim \frac{1}{\Delta p} + G \Delta p. \tag{7.8}$$

The second result is more recent [Mag93], and stems from a thought experiment that embeds Heisenberg's microscope in the vicinity of a black hole and attempts to measure the area of the black hole's horizon. The uncertainty in that case is:

$$\Delta x \gtrsim \frac{\hbar}{\Delta p} + \alpha G \Delta p, \tag{7.9}$$

where α is a constant which the model-independent arguments cannot predict. Here the uncertainty is due to the impossibility of directly measuring a black hole without it emitting a particle that is carrying energy and thereby changing the area of the horizon of the black hole. The smaller the wavelength of the emitted particle, the larger the distortion that it yields.

The theoretical motivation that underlies the two results just mentioned, equations (7.7) and (7.9), was presented in Chapter 5. In the case of String theory this motivation comes from *T-duality*: the increase in energy–momentum of a closed string in the attempt to localize it at smaller and smaller scales increases the string's size. But while the absolute uncertainty in length measurement together with the idea that the string length l_s is the minimum measurable distance were suggested almost a generation ago by String theorists ([Ama89], pp. 111–112; [ACV89], p. 46), a precise theoretical framework for the second term in (7.6) has not yet emerged. I thus agree with Rovelli, who in a recent evaluation of String theory as a theory of quantum gravity, complains that

An actual proof of ultraviolet finiteness is still lacking. At least, I have searched and much asked around, but I have not yet been able to find a reference with such a proof. But String theorists appear to be convinced of finiteness, and I believe them.

([Rov11])

One way to appreciate this problem is to recall that in contrast to EM, QED, and QFT, where the idea of finite extension was introduced to replace point interaction in three-dimensional space (see Chapter 4), the basic building blocks of String theory do not "live" in this space (or even in Minkowski spacetime), but in a higher dimensional space. But the physical localization, or point-interaction, which they presumably come to replace, commonly described operationally in QFT by some causal process, does occur in real space. And so despite the metaphor of the string as an extended object, it remains unclear in what sense this object is actually extended in real space (see [Sch08], Section 6, for a particularly penetrating criticism on this issue). The problem seems to arise from confusing the geometric properties of

String theory with the physical process of localization in spacetime, a process which requires physical access exactly to those degrees of freedom which are, according to String theory, *beyond* real space. Yet this hypothesis, namely the existence of an infinite number of degrees of freedom besides the ones we see, defining extended elementary objects, is at the very basis of String theory, and it alone is responsible for the "stringy ultraviolet finiteness" ([Muk11], p. 8).

String theorists, however, are not only unhelpful in this respect, but seem to insist on the redundancy of physical spacetime altogether:

Once one replaces ordinary Feynman diagrams with stringy ones, one does not really need spacetime anymore; one just needs a two-dimensional field theory describing the propagation of strings. And perhaps more fatefully still, one does not have spacetime anymore, except to the extent that one can extract it from a two-dimensional field theory.

([Wit96], p. 28)

According to our thesis *L*, a charitable reader should interpret this passage along the lines of the weak version of the dynamical approach. Even String theorists cannot but employ spatial (geometrical) concepts such as "length" in their theory (as they actually do). The dynamical (two-dimensional conformal field) theory does not precede geometry in any sense, and so it is doubtful that "one does not really need spacetime anymore." Since geometrical concepts are assumed at the very basis of the theory, at most it gives us a consistency proof for the (low-energy) behavior of geometrical objects in ordinary spacetime. This, I believe, is the only coherent interpretation of statements, commonly made by String theorists, that spacetime is "emergent." We shall say more on this in the final section of this chapter.

Note, again, that thesis *L* implies nothing about the properties of spacetime. The point here is that geometrical notions must be (and, as a matter of fact, are) presupposed, no matter what the geometrical structure of spacetime turns out to be.

7.2.2 Black hole thermodynamics, AdS/CFT, holography, and all that

Despite philosophical doubts about taking thermodynamics too seriously [Cal01], doubts that go as far back as Maxwell and his famous demon [HS10], most physicists follow Eddington in their belief that the second law of thermodynamics must be universal:

If someone points out to you that your pet theory of the universe is in disagreement with Maxwell's equations – then so much the worse for Maxwell's equations. If it is found to be contradicted by observation – well these experimentalists do bungle things sometimes. But if your theory is found to be against the second law of thermodynamics I can give you no hope; there is nothing for it but to collapse in deepest humiliation.

([Edd28], p. 74)

Circular attempts, for example, [Ben03], to argue for the universality of the second law and for its applicability at the microscale by assuming it from the outset are well known. Relevant to our discussion here, however, are similar suggestions to extend the applicability of thermodynamics, and in particular the applicability of the second law, far into the cosmological scale, to objects such as black holes.

The famous result in this context is the Bekenstein–Hawking proof that the entropy of a black hole is proportional to the area of its event horizon [Bek72, Bek73, Haw75, Wal01].[1] This result serves as a benchmark for all contender theories of quantum gravity as yet another feature that should be reproduced from their first principles (see Chapter 8). The relevance of this result to our discussion is twofold. First, the area of the horizon can be seen as a fundamental geometrical notion that, when equated with entropy, is supposed to represent the number of microstates in the black hole, microstates of the basic building blocks of the presumed quantum theory of gravity. Second, and even more speculative, the thermodynamics of black holes, and in particular the relation between the area of the horizon and the entropy of the black hole, have inspired a group of physicists to propose a new principle, known as the holographic principle, which points at a fundamental discreteness in nature at the Planck scale that limits the amount of information that can be stored in any given spacetime region.

The idea of holography goes back to Gerard t'Hooft ['tH93] who suggested that given any closed surface, one can represent all that physically happens *inside* it by degrees of freedom on the surface itself. The situation can be compared with a hologram of a three-dimensional image on a two-dimensional surface (see Figure 7.2, inspired by [Sus95], p. 6379). The details of the hologram on the surface itself are intricate and contain as much information as is allowed by the finiteness of the wavelength of light or, in our case, the Planck length. The conjecture was further elaborated by Leonard Susskind [Sus95] and Raphael Bousso [Bou02], and is now considered a general guiding principle for any future consistent theory of quantum gravity, as it relates aspects of spacetime geometry to the number of quantum states of matter. The idea has also penetrated what Einstein has called "the secular press," and has sparked the imagination of many science writers (see, e.g., [Bek03, Moy12]).

A further development in String theory, known as the AdS/CFT conjecture [Mal98, Wit98], exemplifies this principle. The conjecture states that a four-dimensional gauge theory is equivalent to a closed String theory (including gravitation) in a particular ten-dimensional spacetime. This correspondence is supposed

---

[1] Bekenstein mentions that the idea that the second law of thermodynamics may break down when a body falls into a black hole was John Wheeler's, see [Bek73], p. 2339 (fn. 19).

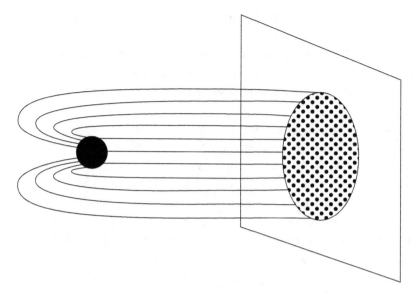

Figure 7.2 Black hole projected on a screen.

to be "holographic" in the sense that gravity degrees of freedom are encoded in a gauge theory that lives in a lower number of dimensions.

The thermodynamics of black holes, the holographic principle, and in particular the AdS/CFT correspondence are, by now, accepted truths among the String community, and yet they exemplify a methodological feature of the theory that, for physicists working in other domains, has always been a serious drawback: (very) nice and presumably consistent mathematical results are, nevertheless, detached from actual empirical verification, and so their role in theory construction is somewhat limited. For this reason I shall restrict the remaining discussion on this subject to two remarks that relate to the insights we have distilled in earlier chapters.

First, whatever discreteness may arise from these results, it is clearly much more complicated and more contrived than the simple (and maybe even simplistic) intuition about discrete spacetime that fueled research in the 1930s and the 1940s. In this respect it seems that not only is Einstein's epistemological argument for thesis *L* vindicated, but so is his worry about the departure from continuum-based physics towards a purely algebraic description of nature (Section 6.5.3). Moreover, it is still far from clear what the basic degrees of freedom of the fundamental theory that obeys the holographic principle are, and, furthermore, how they relate to the notion of "information" which seems to underwrite the entire program.

Second, the holographic principle implies a radical reduction in the number of degrees of freedom we use to describe nature. It exposes QFT, which has degrees of freedom at every point in space, as a highly redundant, effective, description, in which the true number of degrees of freedom is obscured. Consequently, the principle elucidates in part the tension that exists between discreteness and one of the features that characterizes local QFT, namely, locality: the number of degrees of freedom in any local theory is extensive in the *volume*, but the holographic principle dictates that the information content is in correspondence with the *area* of surfaces. As a result, what is regarded as a local interaction in the former could turn out to be a mere appearance, since the additional degrees of freedom which mediate it in QFT are actually physically redundant. Now violations of locality are well known in non-local field theories (Section 4.3.2), and the physical intuition behind their origin is clear. As we shall see below, a clear intuition is also prevalent in the case of approaches to the quantum gravity problem other than String theory. In contrast, I found the discussions within the String community on these issues impenetrable, and so I shall set them aside.

7.3 Background independent strategies

Background independence (BI) is a manifestation of relationalism, a certain philosophical and methodological view that is central to the history of theoretical physics [Bel11]. We have already encountered relationalism in Chapter 5. The history of this view and its key proponents (Leibniz, Berkeley, Mach) is well known (see, e.g., [Bar89]). Whether or not it is vindicated in classical GTR is still a matter of dispute among philosophers and physicists [Fri83, Ear89], but in the context of theories of quantum gravity, relationalism requires – on a minimal interpretation – that the defining equations of a theory will be independent of the actual shape of spacetime and the value of various fields within spacetime, and, in particular, will not refer to a specific coordinate system or a specific metric.[2] Applying this strategy to quantum theories of gravity is not a trivial endeavor, given that conventional QFT is background dependent, and does require a fixed spacetime background.

The status of BI in String theory is still under dispute, and this situation is due in part to the fact that, despite the general opinion to the contrary, the AdS/CFT correspondence – a correspondence that is sometimes touted against the requirement of BI in that theory (as with this correspondence the theory can be formulated as a field theory on a fixed background [BT09]) – is still just a conjecture, and not a theorem! But even if one grants the conjecture and trades it for a theorem, we are still left in the dark with respect to the fundamental degrees of freedom that are encoded in the asymptotic boundary, which are – by definition – outside spacetime ([Smo05], pp. 23–25). The boundary encodes, for example, degrees of freedom

---

[2] The different configurations (or backgrounds) should be obtained as different solutions of the underlying equations.

that represent a reading of a clock, which is not part of the physical system. This formulation cannot be applied to cosmological problems, where the challenge is precisely to formulate a consistent theory of the entire universe as a closed system with an internal clock, and it also falls short of reproducing (in the low energy limit) a theory that does represent a closed system, namely, classical GTR with compact boundary conditions.

While there have been numerous beautiful attempts to construct a fully background independent String theory, these are not yet properly understood [Rov11]. We therefore turn to other approaches to the quantum gravity problem which explicitly admit BI.

7.3.1 Loop quantum gravity

Loop Quantum Gravity (LQG) embarks on a quantization of gravity by constructing suitable variables which have become known as the Ashtekar variables [Ash86]. Like any other contender to a quantum theory of gravity, LQG still lacks experimental confirmation, but what it lacks in experiments it makes up in theoretical framework. In the last two decades it has become an active research program with more than 100 practitioners worldwide, and has established itself as a viable alternative to String theory. Contrary to the latter, LQG does not purport to be a theory of everything, and confines itself merely to a solution of the quantum gravity problem.

Here we shall, again, forego many technical details (see [Rov04, Rov08b] for a comprehensive exposition) and focus instead on the way the notion of fundamental length arises in the theoretical structure.

Faithful to BI as it is, LQG starts from the basic idea that physical quantities measured in real gravitational experiments are always measured with reference to the "place" and at the "time" determined by (quantized) material objects, and not with reference to a fixed coordinate background. These objects – spin networks, graphs with nodes and links among them, where each link and node carry numerical values which represent abstract entities from which certain properties of spacetime can be reconstructed [Smo04] – play the role of material reference systems. To make contact with experiments, a certain dynamical observable is chosen to represent the geometrical notion of area ([Rov93], p. 810):

$$\hat{A}_n|K\rangle = I_n(K)\hbar G/2c^3|K\rangle, \qquad (7.10)$$

where $I(K)$ is a topological theoretical entity that is related to a spatial surface defined by the respective (quantized) matter field (more precisely, the number of times the basic building block of the theory, the loop, crosses the surface), and $|K\rangle$ is a closed graph of the spin network. The magnitude A_n thus describes the observable given by the area of the surfaces of material objects (what follows also applies to the observable that is chosen to represent volume, see [RS95]).

LQG is thus a description of quantum measurements of areas and volumes in slowly varying gravitational fields. Once a measure of area (or volume) is defined, one can inquire as to its structure. A genuine *prediction* of LQG is that the theoretical entities, picked out to designate the primitive measure of area and volume – observables defined on the kinematical Hilbert space of the theory – have discrete spectra. In the case of area, the output of any measurement thereof is an integer multiple of the fundamental area A_0 ([Rov93], p. 811):

$$A_0 = \hbar G/2c^3 = \frac{1}{2}(l_P)^2. \tag{7.11}$$

On this interpretation, it then follows from the quantum mechanical algebraic structure (of the operators representing the observables) that there is a minimal distance between any two spatial points. In other words, in LQG the primitive notion of length turns out also to be minimal [ARS92]. Note, in parentheses, that the elimination of ultraviolet divergences follows elegantly from this prediction, and does not require additional structures as in String theory.

The consequences of this result are far reaching according to the advocates of LQG:

One consequence is that the quanta of the field cannot live in spacetime; they must "build" spacetime themselves . . . We may continue to use the expressions "space" and "time" to indicate aspects of the gravitational field. . . . We are used to this in classical GTR. But in the quantum theory, where the field has quantized "granular" properties and its dynamics is quantized and therefore only probabilistic, most of the "spatial" and "temporal" features of the gravitational field are lost. Therefore for understanding the quantum gravitational field we must abandon some of the emphasis on geometry. . . . The key conceptual difficulty of quantum gravity is therefore to accept the idea that we can do physics in the absence of the familiar stage of space and time. We need to free ourselves from the prejudices associated with the habit of thinking of the world as "inhabiting space" and "evolving in time."

([Rov04], pp. 9–10)

. . . A spin network state does not have a position. It is an abstract graph – not a graph immersed in a spacetime manifold. Only abstract combinatorial relations defining the graph are significant, not its shape or its position in space. In fact, a spin network is not *in* space: it *is* space.

([Rov04], pp. 20–21)

Here also, an interpretation of these statements along the lines of the strong version of the dynamical approach is a gross overkill; the appropriate reading should be along the lines of the weaker version of the dynamical approach we presented in Chapter 6, i.e., arguing that the dynamics only picks up the structural properties of spacetime. The reason is that LQG admits BI, and nothing more; in fact, it keeps in the background all those aspects of space and time that are also kept in the background of classical GTR (topology, dimension, differential structure). But from BI alone (or from relationalism for that matter) it does not follow that dynamics precedes geometry. That LQG admits BI only means that the relation between dynamics and geometry in LQG is exactly the relation between dynamics and geometry in classical GTR, that is, there is no precedence of the former over the latter; rather, they are on a par (to paraphrase John Wheeler, "matter tells space how to curve; space tells matter how to move"). In particular, it is only by interpretation that one associates a certain theoretical (dynamical) entity as geometrical. In this sense, not only does dynamics not precede geometry, but in effect one must employ geometrical concepts *ab initio* in order to identify a dynamical entity (in the case of LQG, the observable A_0) as a geometrical one.

In accord with our thesis L, LQG simply cannot dispense with a primitive notion of geometry; in it one must pick up a theoretical magnitude and designate it as "length" (or, in this case, as "area"), and must do so by stipulation. This, in effect, is precisely what its proponents do:

The purpose of this Letter is to report on the picture of quantum geometry that arises from the use of the loop variables. To explore the geometry nonperturbatively, we must first introduce operators that carry the metric information and regulate them in such a way that the final operators do not depend on any background structure introduced in the regularization. We will show that such operators do exist and that they are finite without renormalization. Using these operators, we seek nonperturbative states which can approximate a given classical geometry up [to] terms $O(l_P/L)$, where l_P is the Planck length and L is a macroscopic length scale, lengths being defined by the given metric.

([ARS92], p. 237)

We propose here a concrete physical interpretation of our result. We wish to emphasize that we propose it just as a possible interpretation, and not as the only possible one.... our model can be seen as an approximate description of the gravitational field in regimes in which the time variation of the field is slow with respect to the time we are interested in. The gravitational field is coupled to a matter field that mimics the fact that we may identify "objects" in certain configurations of a field theory.

([Rov93], pp. 811–812)

The crucial point, in so far as our thesis L on the primacy of geometrical notions is concerned, is that also in LQG a primitive notion of length is presupposed and not derived, i.e., a certain theoretical magnitude is picked out as designating it. Instead of a small rod (as in Swann's proposal in Section 6.5.2) we have here a very small world-sheet associated with area. This association can be justified only if one can make contact by experiments between the notion of length introduced and the results of experiments given in geometrical terms of "segments," for example, distance, velocity, etc., as measured by means of rods and clocks. Here also, there is no question of deriving the primitive geometrical notions from the dynamics, only of making the dynamics consistent with what we can measure, i.e., with primitive geometrical notions.

One needs to be reminded that it is disputed in the literature on quantum gravity whether the quantum mechanical operators in question truly refer to physical or geometrical magnitudes at all, as these operators are not gauge-invariant (see [DT09], and [Rov07] for a reply). I do not take a stance on this issue here. However, I note that this debate only strengthens thesis L, for the following reason.

The main idea behind this dispute is to designate some degrees of freedom as "clocks and rods," and then to express other degrees of freedom in the coordinates defined by these "clocks and rods." The question of gauge invariance is whether these two sets of degrees of freedom transform in the same way. But this question can only be answered after one designates a theoretical magnitude as geometrical that could serve as reference for a transformation of this sort. So here also we see that the designation of a theoretical magnitude as geometrical precedes dynamical considerations. These considerations, as was said before, can only give us consistency proofs for the geometrical interpretation and the dynamics.

Another, related, point is that it is not clear whether the quantum operators carry any new experimental consequences at the macroscopic scale, in particular whether the quantum mechanical minimal length implies the breakdown (or deformation) of exact Lorentz invariance in the extremely high energy regime (more on that in Chapter 8). Some proponents of this approach, for example, [ACS09], believe that this discreteness, via the deformation of Lorentz invariance, may have phenomenological consequences. Others, for example, [RS03, Hos07], are more skeptical. In any case, the argument for thesis L is independent of the way this issue will be settled.

7.3.2 *The causal set approach*

The causal set approach (for a review see, e.g., [Hen10]) is an offspring of the "path integral" or the sum-over-histories type approaches, in which a space of histories is given, and an amplitude (or more generally a quantum measure) is assigned to sets of these histories, defining a quantum theory in analogy with Feynman's path integrals. This view stems from two kinematic axioms which encode two basic ideas: (1) that the quantum analog of spacetime is essentially discrete and (2) that it is structured by primitive causal relations. As in all other approaches to the quantum gravity problem, the hope is to show that at appropriately large scales, this structure approximates the continuous spacetime structure that characterizes GTR.

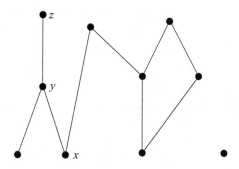

Figure 7.3 A causal set.

The basic principles of the causal set approach are thus twofold. First, a history of the universe consists of nothing but a causal set so that the fundamental events have no other properties except their mutual causal relations (see Figure 7.3, [Hen10], p. 396, where the elements of the set are represented by dots, and the causal relations not implied by transitivity are drawn as lines). Second, the quantum dynamics is defined by assigning to each history a (complex) amplitude. The guiding principle for approximating classical GTR is that for each event in the causal set there will be a respective event in the classical metric manifold such that (a) the causal relations between events are preserved, and (b) there is on average one event from the causal set per Planck volume of the classical metric manifold [BLMS87].

That the causal set approach vindicates thesis *L* about the primacy of geometry is even more evident than in the former two approaches we have discussed so far. This approach purports to recover the geometry of spacetime from a more fundamental causal structure, namely, an order relation between events. From the text below ([BLMS87], p. 521, emphasis mine), it appears that this recovery is meant to be in the stronger sense we distilled in Section 6.5.2:

Now a partial ordering is a very simple thing, and it is natural to guess that in reality g_{ab} [the metric – AH] should be *derived* from *P* [the order relation – AH] rather than the other way around.

The proponents of this approach admit, however, that the partial order is insufficient to achieve this task ([BLMS87], p. 522):

The problem with this is that *P* lacks the information needed to determine the conformal factor $|\det g|^{1/n}$. In a manner of speaking, we get from *P* the metric, but without its associated measure of length (or better, volume).

Their solution is telling ([BLMS87], p. 522):

There seems to be no way to overcome this problem within the context of continuous spacetime, but on the other hand there are many reasons to doubt that spacetime is truly continuous, including of course the infinities of quantum field theory and the singularities of general relativity. If instead we postulate that a finite volume of spacetime contains only a (large but) finite number of elements, then we can – as Riemann suggested – measure its size by counting. If this is correct, then when we measure the volume of a region of spacetime, we are merely indirectly counting the number of "point events" it contains. No attempt to "pack more points into the same volume" could change their density, because it would only increase the physical volume of the region in which they were placed.

As can be seen from the quote above, [BLMS87] aim to derive the geometry from the order relation between the elements of the causal set. But the point is that, as in all other dynamical approaches we have touched upon, it is simply misleading to think here about the notion of "volume" as emerging in any sense from the dynamics. The proponents of the causal set approach acknowledge that in the continuous case there is no unique way of imposing a measure (or a metric) on the space of point-events. But in the discrete case they seem to think that counting the number of events would give uniquely the notion of volume. This move to the discrete case, however, would not help, for the following reason.

To start with, the order relation does not fix the conformal factor – the order relation does not tell us how many elements one can pack into a unit volume – and so the structure that emerges is still without a scale. Acknowledging this, [BLMS87] nevertheless argue that one still gets from this relation the Riemannian metric if one counts events *uniformly*. That is, one gets the Riemannian metric only if the number of elements one packs into a unit volume is invariably the same everywhere. Note that the decision we have here is not how many elements one packs into a unit volume (i.e., it is not the scaling issue above), but whether or not one packs the *same* number of elements, no matter what number this is, into a unit volume in deferent regions. And yet, as in the case of the conformal factor, here also the order relation does not tell us how to count.

I grant that counting uniformly is natural, but the crucial point here is that no matter how natural uniform counting may seem, it does not follow in any sense from the causal order.[3] It is indeed remarkable that in the causal set approach one gets the Riemannian metric from the causal order by counting uniformly (up to a conformal factor), but the order of justification is the other way around: from our well-confirmed classical theory we know that we should get the Riemannian metric, and this justifies the fact that in the causal set approach we count uniformly – if we had to count non-uniformly in order to get the Riemannian metric, we would have felt strongly justified to do so!

---

[3] The problem of "counting" the number of elements in a discrete set is analogous to the problem of imposing a measure on a continuous set to determine the size of its subsets. "Counting" the number of elements in a discrete set depends on the way one individuates and aggregates the members of the set. I thank Meir Hemmo for discussions on this point and on the causal set approach in general.

In sum, the way one counts in the causal set approach amounts to an interpretative move, in which one associates a certain theoretical magnitude (in this case, the number of events) with the geometrical notion "the volume of a region." I thus prefer to understand the fact that a geometrical structure "emerges" in some sense from the fundamental theory here as a consistency proof, precisely because, as we said above, the order of justification is reversed: we know from our classical theory and from our experience that the spacetime metric is Riemannian, and it is this fact that justifies our association of the number of elements with the geometrical notion "the volume of a region."

Support for this insight comes from another thorny problem in this approach, dubbed "the inverse problem" [Smo05]: if the causal set theory is supposed to be more fundamental so that one can "derive" spacetime geometry therefrom, then there should be some procedure to define the classical structure of spacetime $\{M, g_{ab}, O_n\}$ (where M is the differential manifold, g_{ab} is the metric, and O_n are all other matter fields) from some kind of low energy limit of the causal set theory. However, while it is easy to construct – in the sense discussed above – a causal set if the classical manifold structure is given in advance, almost no causal set approximates a low dimensional manifold in that sense; worse, we do not have a characterization – dynamical or otherwise – expressed only in terms of causal relations, which will allow us to pick up those sets that do approximate spacetime.

The notion of a causal set, it seems, is simply too weak to carry the weight of spacetime geometry. Over and above thesis L, Einstein's second worry about a departure from continuum physics is, again, fully vindicated.

7.3.3 Kinematical approaches

Both LQG and the causal set approach are background independent dynamical theories in which one (1) postulates some basic combinatorial building blocks (graphs of spin networks, or causal sets), (2) associates these with some geometrical notions of area or volume, and (3) predicts or assumes discreteness of space or spacetime, respectively.

In addition to these theories, there exist several background independent kinematical frameworks which have not yet matured into full blown theories, but which incorporate many ideas on the notion of fundamental length that we have encountered in earlier chapters. Two of these frameworks which merit special attention are Deformed Special Relativity (DSR) and modern non-commutative geometry.

Deformed special relativity

DSR began [AC01, AC02c] as a revolutionary idea that is based on an analogy of the transition from Galilean relativity to STR: one could describe the transition from

Galilean spacetime to Minkowski spacetime as a consequence of the introduction of an observer-independent velocity scale, namely c. Once we accept c as such (by elevating the empirical result that the velocity of light is independent of the velocity of its source to the level of a principle, namely the "the light postulate" of STR), the kinematical structure employed by the theory changes, and with it so does the symmetry group that characterizes this structure.[4]

Consequently, the simple velocity addition law in Galilean relativity (which involves no velocity scale) $v' = v_0 + v$ is "deformed" and is replaced with the velocity addition law that takes c into account: $v' = (v_0 + v)/(1 + v_0 v/c^2)$. What this means is that even if the mathematical structure that best characterizes our spacetime is Minkowskian, at small enough velocities we could still approximate it as Galilean, while keeping in mind, of course, that there are no preferred frames (i.e., that simultaneity is still relative).

DSR involves a similar kinematical shift. It introduces yet another observer-independent scale, namely a fundamental length scale of the order of the Planck length. By accepting this length scale, we could still maintain the principle of relativity (i.e., that the laws of physics take the same form in all inertial frames) and just modify the transformation rules between frames to preserve the new scale (as we did in the shift from Galilean to Lorentz invariance). Once more, this means that while the mathematical structure that best describes our space is discrete, at low enough energies (or large enough wavelengths) we could still approximate it as continuous. In order to accommodate a fundamental, minimal, length (or, equivalently, a maximum energy) as an observer-independent scale in the theory, DSR suggests a non-linear modification of the action of the Lorentz group on momentum space. As one would expect, the consequences of this deformation for conservation and composition laws are highly non-trivial [JV03, MS03, GL05, Hos07].

According to DSR, STR (characterized by the symmetry group of Minkowski spacetime or, equivalently, by the Lorentz covariance of the fundamental non-gravitational interactions) is thus only an approximate theory and may not be applicable when (quantum) gravitational interactions are present. What DSR offers is a modification of STR that is not committed to a preferred frame, and links deviations from exact Lorentz invariance to the existence of a fundamental minimal length by introducing the latter as an additional observer-independent scale in the kinematical structure, over and above c (the velocity of light).

At this stage there are several strategies for translating the kinematical structure of DSR, formulated in momentum space, into a consistent field-theoretic framework

---

[4] Recall that such an intuition also motivated Born and Infeld [BI34] in their non-linear modification of Maxwell's electrodynamics, see Section 4.2.2.

that can generate testable predictions. All these strategies agree on the starting point, namely a Planck scale modification of the dispersion relation between energy and momentum, from

$$E^2 = c^2 p^2 + m^2 c^4, \tag{7.12}$$

to

$$E^2 = c^2 p^2 + m^2 c^4 + f(E, p, m; l_P), \tag{7.13}$$

just as the atomic structure of matter modifies continuum dispersion relations once the wavelength becomes comparable to the lattice size.

From a phenomenological perspective, different views of DSR disagree on the possible consequences of these modifications, for example, possible variations in the velocities of massless particles, and, if so, whether these variations depend on energy or on energy density (see [AC06] for the former, and [Hos07] for the latter). The physical intuition here is that a localized highly energetic particle produces a back-reaction on the metric, which in turn affects the dynamics of the propagating particle [Rov08a].[5] While both these approaches lead to *in principle* testable predictions, to be compared with forthcoming data from gamma ray bursts (observed by the FERMI telescope), the former proves testable in practice while the latter does not. Be that as it may, since these predictions are different from a straightforward breakdown of Lorentz invariance [GP99] (and at least in the former case, also from exactness thereof [Hos07]), proponents of DSR [Smo10] see it as a promising route to the much sought for "quantum gravity phenomenology." We shall say more on this in Chapter 8 below.

From a broader perspective, one may ask what role DSR should play in the quantum gravity research program. While motivated by LQG, no rigorous derivation of DSR from LQG exists to date. Here I suggest a possible conceptual link that relies on the analogy with STR.

STR can be described as a kinematical constraint on the non-gravitational interactions, quantum-field-theoretic laws being Lorenz covariant. It is thus plausible to assign DSR a similar role in constraining the dynamics of future theories of quantum gravity which extend QFT interactions to include gravity. The analogy fits well with one of the major requirements of DSR, namely the requirement to approximate STR (and the Lorentz transformations) in regimes where gravitational interactions can be neglected.

Compelling as it may seem, however, this analogy must be handled with care. Since all observers agree on the physics once the fundamental length scale is

---

[5] A more down to earth picture is offered by Sabine Hossenfelder, a quantum gravity phenomenologist, on her popular blog (see Section 8.2): think of a family trying to get through a crowd. The kids, small as they are, would arrive before their larger parents.

factored into their measurements, DSR establishes a deformation of STR without giving up the principle of relativity itself. But how, in the framework of this theory, can we probe the ontological claim that space is fundamentally discrete, or that the Planck length is physically meaningful? After all, the fact that by combining other physical scales we can construct l_P with dimensions of a length is immaterial to its meaningfulness. In particular, while c enjoys a robust body of data suggesting its physical interpretation as the speed of light, so far we have no hint on the physical interpretation of l_P.

This dis-analogy underlines what some see as the main problem with DSR, namely the lack of physical motivation for deforming STR over and above the theoretical intuition regarding fundamental length and the hope to render quantum theories of gravity testable. Here one should note that in its early days DSR did in fact enjoy a motivation of this sort: it was expected [AC03a] to explain anomalous observations in energy thresholds of cosmic rays (the anomaly, known as ultra-high energy cosmic rays, or "the GZK paradox," was first observed in [Lin63]). Discouragingly for its followers, however, it is now acknowledged that DSR cannot (and does not) predict these anomalies [AC07a]. Others try to motivate the theory from cosmological considerations [Mag03], but these attempts also lack an empirical underpinning. On the other hand, STR has been tested again and again, and still remains "in remarkably rude health" ([AC07b], p. 802).

The absence of empirical motivation is not unique to DSR; all competing approaches to the quantum gravity problem currently share this feature, and so far have confined themselves to consistency proofs of their respective structures with verified features of classical spacetime. DSR, however, makes a commendable move in that it goes beyond mere consistency, striving as it does to extract new phenomenological consequences from spatial discreteness (see Chapter 8).

Non-commutative geometry

Non-commutative geometry is a modification of quantum field theory, obtained by taking the position coordinates to be non-commuting variables:

$$[x^i, x^j] = i\theta^{ij}, \tag{7.14}$$

where θ is an antisymmetric (constant) tensor of dimension length squared, known as the deformation parameter (in the limit $\theta_{ij} \to 0$ one obtains ordinary spacetime). The formal role of this parameter is much like $\hbar$ in the commutation relation between position and momentum; its physical interpretation is that of a smallest observable area in the ij-plane that can be resolved via interactions. The non-commutativity in equation (7.14) leads to a position space uncertainty principle, which excludes the possibility of localized field configurations, and signifies a minimal uncertainty among spatial coordinates of the form

$$\Delta x_i \Delta x_j \gtrsim \frac{1}{2} |\theta^{ij}|. \tag{7.15}$$

It should be stressed, however, that as is the case with the notion of absolute uncertainty, introduced already in the 1930s, the non-commutativity under discussion here is not inherently quantum mechanical. Rather, it is a formal device used to represent a particular class of interactions between fields, which can exist in either classical or quantum field theory.

Since the inception of these ideas by Heisenberg, Born, and Wataghin, and the elaboration thereof by Snyder and Rosen (Chapter 5), the class of theories that incorporate non-commutative geometry has not been studied seriously, except by mathematicians (e.g., [Con94]). The reason seems obvious, as these theories break relativistic causal order, are inherently non-local, and their implications for the status of Lorentz invariance are still unclear (see below, Chapter 8). In the last two decades, however, with the growing understanding that the future theory of quantum gravity would have to come to terms with these difficulties, theoretical physics has been showing a renewed interest in them (for a review see, e.g., [DN01]).

The non-commutativity can be extended from the spatial coordinates to the algebra of functions thereof $f(x)$, where the product of such functions is known as "the star product," or the "Weyl–Moyal product" [Moy49]. This product, of key importance to non-commutative geometries, is relevant to our discussion on minimal length since it allows us to represent it not only as arising from the non-vanishing commutator in (7.15), which already introduces discreteness in the guise of a finite measurement resolution, but also from a Gaussian distribution in position space (used, e.g., in [Edd46, Ros47, Ros84]) centered around 0, with a spread σ,

$$\Psi_\sigma(x) = \frac{1}{\pi\sigma} \exp\left(-\frac{x^2}{\sigma^2}\right). \tag{7.16}$$

When one works out the star product of two such Gaussians with spreads $\sigma_1, \sigma_2 < \theta$ (for a concise exposition see [Hos13], pp. 43–44), the width of the product is larger than θ. Consequently, since Gaussians with smaller width than θ have the effect of spreading, rather than focusing, the product, one can think of the Gaussian with width θ as having a minimum effective size.

Ideas germane to non-commutative geometry appear also in DSR and the attempts to deform special relativity [FKGN07], and in recent attempts to develop further a proposal that goes back to Born and Wataghin (Sections 5.2.3, 5.3.2) to depict physics as unfolding on a curved momentum space [KGN03]. We shall say more on these attempts below in Section 7.5 and Chapter 8.

7.4 Emergent gravity

A broad class of approaches to quantum gravity adopt a condensed matter perspective from which gravity-like features "emerge" (in different degrees of refinement and complexity) in various models and in specific regimes, by way of reorganizing the degrees of freedom of a certain underlying model in a way that leads to a regime in which the relevant degrees of freedom are (1) qualitatively different from the microscopic ones, and (2) have gravitational characteristics. In this approach, gravitational dynamics is recovered at the macroscopic scale, more or less successfully, from an underlying (microscopic) non-gravitational system. A widely quoted example in this context is Unruh's [Unr81] discovery that in some situations the properties of sound propagation in a moving fluid can be exactly described as the propagation of waves in curved spacetime with a pseudo-Riemannian metric. The general idea behind this fact is that, when a fluid is flowing with a given velocity field, sound waves are dragged by the fluid flow, so if a supersonic flow is realized in some region, sound waves cannot travel upstream. The phenomenon is analogous to what happens in general relativistic spacetime regions that trap light, and so can serve as a sonic analog for black holes [Unr95].

There are obvious methodological advantages to this approach. Quantum gravity and, maybe less questionably, cosmology, are commonly thought to be non-experimental disciplines that involve systems that can only be studied observationally on cosmological scales. With analogous models of gravitational phenomena [BLV11], however, one could study the physics on effective curved spacetimes in systems which could be realized experimentally in a laboratory. This, along with the insights that can be gained for the study of the relationship between spacetime geometry and the more fundamental structures that are postulated by other approaches to the quantum gravity problem, may help make the detection and the testing of some features of QFT on curved spacetimes a real possibility.

The research program of Emergent Gravity seems to suggest that notwithstanding the absence of anything resembling spacetime at that microscale, certain features of spacetime geometry "emerge" at the macroscale (for a recent analysis of this claim see [Bai13]). And yet, if only for reasons of suggesting possible empirical verification of the underlying fundamental theory, here also one must point at a certain energy scale in which the basic building blocks, the non-gravitational degrees of freedom, make themselves noticeable, and so our thesis L seems to be, again, vindicated: no matter how "non-geometrical" these basic building blocks are, one must still pick a set of them as designating some geometrical notion of length or volume to make contact with experimental results.

What about minimal length? Are theories of Emergent Gravity introducing discreteness to spacetime? Unfortunately, as far as this notion is concerned, there is no general statement that can be made about its existence using this approach. What is true is that the putative limit on the spatial resolution of structures cannot be deduced from the distinction between macro and micro alone; as is well known from statistical mechanics, coarse graining by itself does not imply finite measurement resolution, unless one introduces additional counting arguments in the attempt to connect microelements with macroscopic properties.

7.5 On the "disappearance" of spacetime

At the time this manuscript is being written, the LHC has found evidence of the Higgs Boson, and QFT, the theory behind the Standard Model, has won another stunning confirmation. High as the energy levels we have been able to probe are, we are still very far – almost 16 orders of magnitude – from probing those levels which are pertinent to the quantum gravity problem. Nevertheless, proponents of the theories that purport to solve this problem have already made some astonishing claims regarding their philosophical consequences, in particular with respect to the notions of space, time, and spacetime, or rather, the disappearance thereof. In each of these cases, a certain feature, deemed essential for spacetime to carry ontological weight, is challenged as non-existent in the fundamental theory.

Arguments for the "disappearance of time" have been voiced by physicists ever since GTR was rewritten in a constrained Hamiltonian form [Ber49, Ear03b]. They became popular partly due to John Wheeler's ideas about quantum foam ("no before and no after" [FW98], p. 247), and have recently resurfaced in the debate over the so called "problem of time" in quantum gravity [BI99].

The debate is old, and involves applications to modern physics of various meta-physical opposites – ancient Greek (e.g., Parmenides versus Heraclitus [EB99]), early modern (e.g., Berkeley versus Newton [Ear89]), and contemporary (e.g., A versus B theories of time [Ear02]). The main drive behind these eliminative arguments is a relational reading of the principles of equivalence and general covariance: once one accepts that the spacetime metric is dynamic (i.e., interacts with the other matter fields), and that there is no preferred foliation on the spacetime manifold, one must admit, or so the story goes, that, when treated globally, the universe as a whole is a block universe, i.e., a physical system whose energy is constant, in which no change occurs, and whose physically meaningful magnitudes are "frozen." And since time is commonly thought of as the measure of change (for a unique view to the contrary see [Sho69]), without change there seems to be no place for time in quantum gravity.[6]

---

[6] Additional support for this thesis comes from the fact that a central equation in canonical quantum gravity, the Wheeler–DeWitt equation, has no time parameter in it.

Quantum gravity not only has no place for time, it also has no room for space. The relational reading of the principles of equivalence and general covariance also motivates – as we have seen – the arguments for the disappearance of the latter. In this case, it is not change which is deemed essential but non-existent, rather it is geometry: once dynamical, non-geometrical, degrees of freedom (causal sets, spin networks) presumably take precedence over geometry, the latter is supposed to "emerge" from the former without any geometrical input. But without geometry what role is left for spacetime in the fundamental theory?

Relationalism, in short, motivates both the disappearance of time and the disappearance of space as fundamental physical entities, by eliminating change and geometry, respectively, from theories of quantum gravity.

Our thesis L, however, should give those who promote elimination of this sort a reason to pause. This chapter has hopefully convinced the reader that geometry is here to stay in quantum gravity. Thesis L applies also to the argument for the disappearance of time since duration and change in a block universe can be (and in fact are) described, albeit locally, or internally, when one picks a subsystem of the universe and designates it as a clock, thereby depicting subsequent local evolution of all other systems with reference to that clock. Agreed, this designation only forfeits claims for the elimination of geometrical notions such as length and duration, but given the additional (weak) premise (accepted implicitly by proponents of the elimination schemes), that these are essential for a certain structure to be considered spatiotemporal (otherwise the elimination argument would not succeed), it is sufficient as a demonstration that relationalism cannot go beyond its main premise: while spacetime may well be nothing more than the relations between dynamical objects, it is certainly nothing less.

Relationalism, however, is not the only motivation for the arguments for the disappearance of spacetime in theories of quantum gravity. Another route for this astonishing conclusion stems from the discreteness of the primitive notion of length that these theories assume or predict. Here, what seems to be essential for spacetime is not geometry *simpliciter*, but a continuous one at that. Let us focus on a typical statement to this end, made in the context of the modified uncertainty relations and the absolute uncertainty in length measurements:

In view of the absolute minimum position uncertainty one may plausibly question whether any theory based on shorter distances, such as a spacetime continuum, really makes sense. Indeed in light of the fact that laboratory experiments which probe small distance properties of particles are all high energy scattering experiments, one might conclude that spacetime at such small scales may not be a useful concept, and that spacetime at the Planck scale may not even exist in any meaningful operational sense. Such ideas are not new and were espoused in the era of S-matrix theory in the 1960s, that is that the scattering amplitude expressed in terms of input and output momenta may be the fundamental reality of high energy physics, and not point-like or string-like particles in a spacetime continuum. . . . One might even speculate that the spacetime continuum concept actually impedes physics in the same way that the concept of an ether impeded physics in the 19th century. As such, a theoretical structure based entirely on momenta, such as a modern version of S-matrix theory, might be desirable and interesting.

([AS99], p. 7)

Contrary to the former line of thought, which is driven by relationalism, this argument accepts that a primitive notion of "segment" exists, and therefore does not target geometry as an essential feature of spacetime that is to be eliminated. It does, however, draw a far reaching conclusion from the discrete nature of this geometry. But here, again, one should be careful. That spacetime ceases to exist as a fundamental entity at the Planck scale follows from this argument only if one accepts the additional premise that the continuum is an essential property of spacetime. The lesson we have learned from the philosophy of mathematics (Chapter 2) is that there are no good reasons to accept this premise.

There is, however, a subtle point here that merits further attention. Even if one accepts that geometry is on a par with dynamics, and that spacetime does not have to be continuous in order to meaningfully "exist," so that alternative, discrete, structures can carry an ontological weight, one may still be puzzled by the special role played by momentum space in theories of quantum gravity. Historically, this special role of momentum space as a legitimate representation of physics can be traced, as we have seen, back to von Neumann, Wataghin, and Born (see Sections 4.3.2 and 5.2.3). This representation still requires a geometry and, in the cases under consideration here, this geometry may even be curved, and yet, as argued in the above quote, it appears to be more appropriate as a description of reality at the Planck scale than in position space. In particular, familiar spatiotemporal notions such as "locality" may now become secondary, dependent as they are on specific features of the geometry of momentum space.

7.5.1 The case of locality

Let us exemplify this curious situation with the status of locality in theories such as DSR, that regard momentum space representation as more fundamental than position space.[7] The fact that the Planck length acts as a regulator in the ultraviolet is manifest in the introduction of the corresponding Planck mass as an observer-independent energy scale in momentum space. The requirement that

---

[7] One should be reminded that – natural as it may be – the momentum space representation of DSR is not problem-free. For example, the fact that the momentum of free particles becomes a non-additive quantity leads to conceptual problems in the formulation of a field theory, e.g., the proper definition of conserved quantities in interactions, or the transformation of multi-particle states, see [Hos07] for an analysis.

Lorentz transformations in that space leave this scale invariant leads to a class of deformations of STR in position space [AC01, AC02c]. So far, there is no agreed way of translating DSR from momentum space to position space. This translation is crucial not only to the philosophical question we are interested in here, but also for reasons of overall consistency with standard QFT and with possible observations that can test DSR: although we can derive the equations of motion in momentum space without actually knowing the transformations in position space, we should still look for a position space formulation of DSR, given the intimate connection between spatiotemporal symmetries and conservation rules, and the hope to verify experimentally claims such as an energy-dependent velocity of light in recent data accumulated from distant gamma ray bursts [ACS09].

A possible translation for a single free particle exists which leaves intact the Lorentz transformations (and the constancy of c) while keeping the minimal (proper) length observer-independent [Hos06]. The problem, however, arises when one introduces interactions, or when one considers multi-particle states. Here, the deformations of Lorentz transformation inevitably lead, via the modifications of the dispersion relations, to an energy (or an energy–density) dependent speed of light and, as has been argued [SU03, Hos09], to loss of locality.

The gist of the problem is rather simple. Consider a case where the velocity of light decreases monotonically as the energy increases. When the energy saturates the Planck mass bound, the speed vanishes; a photon with $E = m_P$ would thus be at rest. Suppose we put this photon inside a classical (macroscopic) box in which modifications of STR or GTR are absent or at least negligible. Clearly, there are frames for which the photon and the box are at rest, but there are also frames in which the box, suffering no Planck scale modifications, is moving with a certain velocity while the photon, having a frame-independent mass, is still at rest! In these frames the photon cannot remain inside the box, and if we wait long enough we should find the photon arbitrarily far from it ([Hos10] calculates the discrepancy for a generic macroscopic set-up).

Things become even weirder if one brings interactions into the picture, and inquires about the point-coincidence of events, since the different transformation behavior of the world-lines of the box and the photon results in an observer-dependent notion of what constitutes the same spacetime event. Much as the introduction of an observer-independent constant (c) deformed the Galilean transformations and yielded the relativity of simultaneity, the introduction of another observer-independent constant (m_P) deforms the Lorentz transformations and yields what has been recently dubbed "the relativity of locality" [ACFKGS11, AC12a]. The idea is interesting for its own sake, but is especially relevant to our discussion with regard to the claims for the disappearance of spacetime, as it leads one to abandon what Einstein saw as the most basic principle that underlies a spatiotemporal description, namely, the principle of point-coincidence. This principle, along with the principle of relativity and the principle of the constancy of the velocity of light, underwrite time dilation, length contraction, and the relativity of simultaneity. It ensures, in other words, that, although we use differently calibrated rods and clocks, we all still live in the *same* spacetime.

In a nutshell, relative locality amounts to the following. In the low energy limit of STR and GTR, where $m/m_P \to 0$, two observers may label the same event with different spatiotemporal coordinates, but if two events coincide both in space and in time, i.e., occur at the same point in spacetime, both observers will agree on their coincidence, regardless of how they measure their (proper) time and their (proper) distance therefrom. Not so in the high energy scale of DSR, where $m/m_P \to 1$. Here two distant observers would disagree on whether or not two events occurred at the same point. What determines their disagreement is the geometry of the (curved) momentum space: when momentum space is flat, STR, as well as absolute locality, are recovered.[8] In other words, different observers, separated from each other by translations, construct different spacetime projections from the invariant curved momentum space. Nonetheless, all observers would agree that interactions are local in the spacetime coordinates constructed by observers local to them, which means that the non-locality induced by DSR is just a coordinate artifact.

In this extreme operationalist picture, spacetime and energy and momentum have become intertwined [KMM04]. Even ignoring gravitation, one can no longer talk about position and time without a material object being "there" and "then," to the extent that the properties of this object's position and time now depend on its energy. At low energies, the dependence is very weak, so we have the illusion that spacetime exists independently of the (test) particles that might "fill" it. But this illusion breaks down at high energies.

In the previous arguments for the disappearance of spacetime, either geometry itself, or a certain feature thereof, were claimed to be absent at the Planck scale. In the case under consideration here, yet another feature – point-coincidence – is targeted as non-fundamental. It may thus appear that an argument for the disappearance of spacetime based on relative locality may succeed where its predecessors failed, for no matter how one insists on the primacy of geometrical notions, or on the possibility of alternative structures for the continuum, the fact (if it is a fact) that locality is no longer absolute means that we do not live in the same spacetime. Put differently, it is only after one specifies the curvature of momentum space that different observers can decide (and measure) the behavior of objects in spacetime.

But here one needs to tread carefully. What spacetime we live in may depend on the curvature of momentum space, but this curvature is an empirical matter

---

[8] For possible experimental set-ups that can measure this curvature see [ACFKGS11], pp. 8–9.

of fact, to be decided with experiments done in spacetime, and so the argument from relative locality (or from the primacy of momentum space representation) to the disappearance of spacetime succumbs to thesis *L* as much as the relational arguments about the disappearance of space or the disappearance of time do.

For this reason, the best way to interpret the argument from relative locality, I believe, is again along the lines of the weaker version of the dynamical approach we have distinguished above in Section 6.5.2: there is no sense in which the dependence of point-coincidence on the curvature of momentum space could be interpreted as an eliminative reduction. The lesson of thesis *L* is that no eliminative reduction of spatiotemporal concepts to dynamical ones is possible, if the dynamical theory that is supposed to carry this reduction purports to be empirically verifiable. What the argument from relative locality does suggest, I think, is that different features of spacetime may depend on more fundamental features. Under this weaker version of the dynamical approach, no attempt is made to argue that spacetime is less fundamental, but rather what is shown is how one of its features supervenes on a more fundamental feature such as the curvature of momentum space.

7.5.2 A final word

We have seen that in theories of quantum gravity, at least those which have matured so far, a certain dynamical entity is always picked up to designate geometrical notions (e.g., length, area, or volume). In this sense, spacetime can only "emerge" at the Planck scale if some geometry is presupposed *ab initio*, and therefore cannot be eliminated. We have also seen that attempts to eliminate spacetime which are based on the idea that at the Planck scale geometry is discrete, at most succeed in showing that certain features of spacetime, for example, the principle of point-coincidence, or the appropriate symmetry group that characterizes spacetime, depend on more fundamental features, such as the curvature of momentum space. Consequently, spacetime and the geometrical concepts it rests on (length, area, volume, time-interval) all remain untouched by these arguments.

The claim that spacetime ceases to be fundamental at the Planck scale can only be interpreted along the lines of the weak version of the dynamical approach, namely, (1) that the properties or symmetries of spacetime depend on some dynamical entities, or (2) that certain features of spacetime depend on features which are more fundamental. Moreover, the dependence of spacetime features (in this restricted, weak, sense) at the Planck scale on dynamical features of the underlying theory, is not a result of a philosophical thesis (relationalism), but a consequence of an additional, empirical, hypothesis about minimal length, manifest, for example, in the curvature of momentum space, which is supposed to be of the order of $1/m_P^2$.

This last point reminds us once more that proponents of discrete spacetime must face two equally difficult challenges. First, they must show how the structure they favor approximates the standard continuous structure that physics has been working with for so many generations. But this proof of consistency is not sufficient. For no matter how consistent the discrete structure may be with current empirical findings that also support the continuum picture, it must also predict new phenomena, so that the physics community will take it seriously not only as a viable alternative to the continuum, but as the one structure that actually holds at the Planck scale. This search for quantum gravity phenomenology is the subject of our final chapter.

8

The proof is in the pudding

8.1 Outline

Up until recently it was not uncommon to find skeptical remarks in the literature about the possibility of testing or observing quantum gravitational effects (see, e.g., [Ish95]). As the energy scale of which these effects are believed to become manifest is the Planck scale, almost 16 orders of magnitude above the capacity of our best available particle accelerator, attempts to settle experimentally the question of discrete versus continuous spacetime seemed hopeless. One would have to resort to cosmological observations and hope to find in them traces of the early stages of the universe, but without concrete predictions, one would not know where and what to look for.

This situation is almost the opposite of the state of high energy physics between the 1930s and the 1950s.

Back then, the relativistic quantum mechanical framework with which physicists described nature was more or less agreed upon. Many of them used Dirac's equation as a starting point for calculating particle interactions (see, e.g., [Kra92a]). As data from cosmic rays were accumulating [Cas81], mismatches between theory and observation were interpreted by physicists (e.g., Heisenberg and Wataghin) as evidence for a minimal length in QED, that would also cure the theory of its divergences (see Chapter 4). However, as more particles were being discovered [Gal83, Gal87], and as the renormalization program took over [Rue92], both the "evidence" and the theoretical motivation for minimal length were discarded.

Between the 1930s and the 1950s, therefore, as theory and observation went hand in hand, the case for a fundamental length in QFT became weaker and weaker. As we have seen, it remained at the fringe of theoretical physics for many years, and ideas such as non-local field theories with form factors, non-commutative geometry, absolute uncertainty, and reciprocity between momentum and position spaces, made almost no impact on mainstream physics, only to be rediscovered almost two generations later.

Today, in contrast, the theoretical support for fundamental length – albeit at a scale much smaller than the one that was contemplated in the 1930s – is stronger than ever, flourishing as it does with no experimental data that could back it up. This lack of observational input places the quantum gravity community in an awkward position. Since there are almost no relevant data sources, as soon as the slightest shred of evidence is mentioned by experimentalists, theorists jump and immediately start "playing games" before even looking for more established results, or at least clarifying with the experimentalists the relative validity of the data (see [AC12b] for a recent, and rare, reflection on this issue in the context of the alleged "faster-than-light neutrinos" at the OPERA detector).

For the philosopher of physics this situation supplies fascinating case studies for methodological issues, as it sharpens and amplifies the tug of war between conservatism and innovation that characterizes the progress of science.

Modern physics, via the field concept, has been working with the notion of the continuum for many generations. Based on this concept, the Standard Model and QFT, the theory that underlies it and describes all interactions but gravity, have achieved unprecedented empirical verification. Associated with QFT are several physical principles which have become part and parcel of common knowledge and, more importantly, part and parcel of the tool kit of theory construction. These principles are locality, relativistic causal order, unitarity, and local Lorentz invariance. Any departure from these principles would thus be an important physical discovery, with dramatic theoretical consequences.

Principles such as the four mentioned above constrain quantum gravity theorists in their quest for experimental evidence for fundamental length. On the one hand, they would like to construct a model which *in principle* predicts new phenomena in domains yet to be tested. On the other hand, the model should be consistent with data obtained up until now, i.e., it should not be deemed already false! Ironically, here the challenge to come up with a parameter-free model and with definite predictions *without* the ability of fine-tuning is not a blessing but a curse; the risk is that the model could then be confirmed – or killed – in one stroke.

Focusing on the above four principles, in this chapter we shall inquire about the status of quantum gravity phenomenology, with the aim of extracting answers to two philosophical questions. The first is general, and has to do with the methodology of science and the process of scientific progress. Put bluntly, it asks how far are we willing to trust well-confirmed and empirically well-motivated principles in the construction of models for new phenomena. The second, which connects us to the beginning of this monograph, concerns the issue of fundamental length, and the puzzle we started our journey with, for by now we are finally in a position to ask whether the question about spatial discreteness can be deemed empirical, or whether it must remain metaphysical, forever beyond experimental reach.[1]

8.2 The quest for quantum gravity phenomenology

Gravity is the weakest interaction of all – exactly how weak depends of course on the amount of mass one has at one's disposal – but, in contrast to other interactions, it cannot be shielded from and it couples to everything. This twofold fact explains why, despite its prominent role in our daily lives, terrestrial experiments, aimed at examining physics at short length scales, can nonetheless completely ignore gravitational interactions.[2] This negligibility of gravitational effects has been feeding skepticism about quantum gravity phenomenology, as these effects are expected to become comparable to those of the other interactions only at energy scales close to the Planck scale. At 10^{19} GeV, this domain is out of reach at any LHC experiment, and it is even far above – more than 8 orders of magnitude – the highest energies ever observed in cosmic rays.

The problem is thus one of an unfortunate trade-off: we have a lot of good data about physical interactions in flat spacetime, where the effects of gravity are negligibly small, suppressed as they are by the Planck mass (or the Planck length), and almost no data about physical interactions in a strongly curved spacetime – black hole physics or big bang physics, where these effects are dramatically intensified. The first challenge quantum gravity phenomenologists face, therefore, is to find circumstances, either terrestrial or cosmological, in which these effects are amplified and become observable.

Amplifiers for these effects could be, for example, dimensionless quantities, that increase our sensitivity to gravity in ordinary physics set-ups. One such amplifier is the number of particles. As an example ([AC02b], p. 905), think of the proton lifetime predicted by grand unifying theories (GUTs), which is of order 10^{33} years [N+09]. A situation in which 10^{33} protons are monitored can increase dramatically our sensitivity to proton decay, which is practically unobservable without an amplification of this sort. Similar considerations explain the success of Brownian motion studies a century ago, where the relevant amplifier was the number of collisions. Of course, identifying experiments with this rare quality is only the first step in the process; one should also demonstrate that the putative amplifier is relevant to the

---

[1] Apart from the standard reference to published physics articles, in the next section I rely also on information, expressed over the last 5 years, by Sabine Hossenfelder, a quantum gravity phenomenologist, who runs a popular blog at backreaction.blogspot.com.

[2] Recall Feynman's calculation ([FMW03], p. 11) that the gravitational field contribution to the ground state of a hydrogen atom would change the wave function phase by just 43 arcseconds after 100 times the age of the universe.

quantum gravity effects connected with the granularity of spacetime one is trying to probe.

Another possible class of amplifiers is based on the idea that, even in a flat background, some consequences of the underlying discreteness of spacetime geometry, predicted or assumed in theories of quantum gravity, would make themselves manifest via a breakdown of one or more of the physical principles that, among others, characterize ordinary QFT, namely, locality, relativistic causal order, unitarity, and local Lorentz invariance. This manifestation would then be amplified at cosmological scales, say, by the distance traveled by high energy photons.

This second class of amplifiers saddles quantum gravity phenomenologists with another, more difficult, methodological challenge. The problem is that any amplification that relies on the breakdown of well-confirmed and well-established principles such as the four above, must show itself in a way that will still be consistent with the empirical data that have been supporting these principles up until now with ever more stringent constraints. In other words, not only must one point at the possibility of a theory of quantum gravity yielding new predictions, but one must also demonstrate that this theory is consistent with current data, and is not already false!

Here is where quantum gravity phenomenologists must tread cautiously. As Sabine Hossenfelder, a quantum gravity phenomenologist, admits, most of the plausible models that could yield new predictions are extremely hard to construct, as – lacking any experimental data to guide the theoreticians – they require a careful connection with those existing mathematical constraints which are empirically well confirmed. The implausible models – those which disagree with the well-established principles – are, as one might have guessed, less constrained, and therefore are easier to come up with, but these models are also easier to falsify and, unfortunately, these models are the ones that make the headlines in the public debates on the hypothesis of fundamental length.

As an example, consider a recent claim which was first published in *Nature*, soon after made it into *Physics Today*, and subsequently created a lot of hype in the Blogosphere, that quantum gravity effects may be observable in "table-top" quantum optics experiments that are already within our technological reach ([PVA$^+$12], p. 393). The amplifier in this case, according to the proponents of the experiment, is the mass of the quantum system under consideration.

Now it is true that some quantum gravity effects such as the deformation of Lorentz invariance in, for example, DSR scenarios (see Section 7.3.3) are suppressed by the Planck mass, and so it might appear that one could enhance them by aggregating masses to increase the total mass of the system. However, what the proponents of the quantum optics experiment have failed to appreciate is that these deformations must be done in a consistent way. Just breaking Lorentz invariance by introducing a preferred frame (e.g., [GP99]) will not do, as there are already many strong constraints militating against it.

One can keep the principle of relativity intact by deforming Lorentz invariance on momentum space so that the Planck mass becomes a new observer-independent constant (e.g., [AC01]), but the consequences of this move are that (1) momentum now transforms non-linearly,[3] and (2) higher order terms in the summed momenta become more and more relevant as the mass increases, irrespective of the nature of the object under study and the volume inside which the momentum is accumulated, which already stands in flat contradiction to the well-established verdict of STR at the macroscopic scale.[4]

In order to remove this conflict with observation, an adhoc solution was suggested [MS03], according to which the amplifier must be rescaled proportionally to the number of particles involved, namely, the relevant mass scale, the Planck mass, is rescaled to N times the Planck mass for N particles. This rescaling removes the inconsistency with current empirical data, but, of course, it also renders the purported quantum gravity effect harder to detect. In the proposed experiment, for example, it would suppress the sought for effect roughly by a factor of 10^{10} (the estimated number of photons in [PVA$^+$12], p. 396), and so it seems extremely unlikely that the experimentalists would find anything to write home about. For quantum gravity phenomenologists, however, the contrapositive conclusion is much more important: if nothing new is found, the experiment cannot be used as an argument against a minimal length modification of QFT.[5]

The same worry about the smallness of the quantum gravitational effects arises in the analysis of several thought experiments that were conceived during the debate on the necessity of quantizing the gravitational field (see Section 5.6). For better (or, as I shall argue here, for worse), the quantum gravity community still regards these thought experiments today as motivating the quest for fundamental length. One of the most cited among these experiments is the Eppley and Hannah thought experiment [EH77] that is often referred to as a no-go argument against semiclassical quantum gravity.

In semiclassical quantum gravity (quantized) matter fields are coupled to the classical metric which interacts with their expectation values (see Section 5.6). The aim in [EH77] is to demonstrate that this coupling of quantized and non-quantized systems leads one into trouble. The strategy is to propose a thought experiment that can be done *in principle*, and to derive a dilemma from its possible consequences.

---

[3] Appendix A of [PVA$^+$12], however, introduces a linear sum of momenta.

[4] This is dubbed in the quantum gravity literature as the "soccer ball problem," see, e.g., [Hos07] and references therein.

[5] Hossenfelder was quick to point out the flaw in [PVA$^+$12] and to warn her blog readers against drawing hasty conclusions from a possible failure of the experiment to detect any signature.

The experiment looks simple: one prepares a quantum system with a narrow spread in momentum (but with a large spread in position) and measures its position with a gravitational wave detector. If gravity is not quantized, the gravitational field need not obey Heisenberg's uncertainty relations, and therefore it would be possible to prepare the detector with a narrow spread in momentum and a narrow spread in position. Coupling this detector to the quantum system yields, or so the story goes, two possible, equally dire, results.

First, if the coupling of the quantum system to the gravitational wave detector collapses the state of the latter, one can resolve its position to a precision determined by the short wavelength of the gravitational wave, and one can do so without transferring a large momentum, thus violating Heisenberg's uncertainty principle. Any attempt to keep the uncertainty principle intact would result in a violation of momentum conservation. Second, if the coupling does not collapse the state, one violates the no-signaling constraint (imposed by STR and respected by standard quantum mechanics [GRW80]), as the gravitational wave detector could serve as a tool for distinguishing between components of an entangled state without collapsing it, and if these components are space-like separated, one could know instantaneously whether a measurement on the other component has occurred.

As common in any dilemma, both horns are undesirable, and so the conclusion Eppley and Hannah (and with them a generation of physicists) draw is that the gravitational field must be quantized.

That this conclusion is unwarranted can be shown in many ways. First, one may accept that the argument is valid, and just attack one of its premises, thus exposing it as unsound. This route is taken in [Mat06], where it is shown that the thought experiment on which the argument is based cannot be done even *in principle* in our universe as (1) it requires a basically noiseless system and detectors more massive than we have mass available, and (2) the entire device that is supposed to detect the gravitational wave sits inside its very own black hole.

An even shorter route that requires no calculations whatsoever leads one to reject the argument not as unsound but as circular. The point here is that the argument simply begs the question, aiming as it does to demonstrate that quantum mechanics is universal (hence the gravitational field should be quantized) by assuming (!) that quantum mechanics is universal (hence rejecting the possibility that Heisenberg's uncertainty relation would be violated for certain types of interactions). For what disallows one to take the first horn of the dilemma, and thus to resolve the problem, is the idea that an interaction between a non-entangled system (the gravitational field) and a quantum entangled one cannot be described consistently in the composed system. But why should an interaction of this sort in such a composed system obey quantum mechanics unless quantum mechanics is supposed to be universal to begin with?[6] The same circularity can also be traced to the assumption of momentum conservation in the measurement process, which is, as is well known, still poorly understood even within non-relativistic quantum mechanics (see also Section 8.5).

Another "no-go" argument against semiclassical quantum gravity rests on a different thought experiment which was actually performed [PG81]. Here the idea is to demonstrate that semiclassical gravity leads to contradiction with observation, since it allows an amplification of quantum superpositions in certain situations and these are never observed. The attentive reader may recall that the same line of reasoning was prevalent in the 1950s and the 1960s (see [DR11] and Section 5.6), and was already debunked then. That quantum superpositions decohere is common knowledge today. What is less known is the mechanisms behind decoherence. And yet without this kind of knowledge, one cannot rule out an explanation of the experiment in [PG81] which is fully consistent with the gravitational field's remaining classical (albeit very weak), and the quantum superposition's decohering as a result of some *other* environmental noise.

The upshot of this analysis goes back to Hume and his warning that there are no demonstrations of matters of fact. Instead of being frustrated that semiclassical gravity has spoiled their theoretical motivation for looking for fundamental length by stubbornly refusing experimental falsification, quantum gravity phenomenologists should divert their efforts from trying to "prove" that gravity must be quantized, to searching for the empirical signatures of this quantization. After all, the whole point of the exercise is that the question of spatial discreteness is an empirical question – a question which must be answered not with logical arguments, but with experimental evidence.[7]

If we have learned something from the history of the quest for fundamental length, it is that the battle for spatial discreteness cannot be won easily, and certainly not in one stroke. The opposition must first be convinced that the hypothesis of fundamental length is a viable possibility that cannot be ruled out. This may be done by showing how theories that incorporate spatial discreteness reproduce or approximate – at the appropriate scales – well-known results obtained within the standard, continuum, structure. Only then may one change tactics and focus on finding models of quantum gravity that yield new predictions. In what follows we shall take these two routes in turn.

---

[6] This type of circular reasoning is not uncommon in theoretical physics, especially when one aims to prove *a priori* that some physical principle must be universal. See, e.g., the information-theoretic exorcism of Maxwell's demon aimed at "proving" that the second law of thermodynamics is universal, by assuming it from the outset.

[7] A similar attitude is evident in the recent hype around "firewalls" [AMPS12] as a solution to the black hole information loss paradox (see below), which has a perfectly acceptable solution within semiclassical quantum gravity ([Wal01], pp. 29–30).

8.3 Consistency proofs

Ideally, any contender solution to the quantum gravity problem must, as a minimal requirement, reproduce or approximate the well-established and well-confirmed predictions of classical general relativity. In addition, since general relativity is also approximated in flat spacetime by the special theory of relativity, the principles that define STR should also emerge at an appropriate scale from the quantum theory of gravity.

"Ideally," because contenders to the quantum gravity problem have not reached this mature stage yet, at least with respect to the first challenge. Currently no attempt is made by these theories to reproduce, say, the deflection of light (famously tested in 1919), the Shapiro time delay of light, or the perihelion shift of Mercury (for these and other experimental tests of relativity see, e.g., [Wil06]). As a more modest benchmark, research has focused on showing consistency with several features of general relativistic spacetime and with the defining principles of STR (for a comprehensive list of these achievements see Smolin [Smo10]).

We shall set aside the discussion of the second challenge to the next section. Here we shall look instead into three recent accomplishments of theories that introduce spatial discreteness, namely, the calculations of (1) the black hole area law in the thermodynamics of black holes, (2) the number of classical spacetime dimensions, and (3) the positive cosmological constant.

8.3.1 The thermodynamics of black holes

At the purely classical level, black holes in general relativity obey certain laws which bear a remarkable mathematical resemblance to the ordinary laws of thermodynamics (for a comprehensive review see [Wal01]).

First, similarly to thermodynamic entropy, the surface area of the event horizon of a black hole can never decrease with time. Together with the idea that information is irretrievably lost when a body falls into a black hole – an idea which seems to go back to John Wheeler (see [Bek73], p. 2339 and Section 7.2.2) – this fact, derivable solely from differential geometry considerations, and first discovered by Hawking [Haw71], led Bekenstein to propose [Bek72] the generalized second law, where the area of the event horizon of a black hole (multiplied by some factor) is interpreted as its entropy, so that the sum of the ordinary entropy of matter outside the black hole, plus a suitable multiple of the area of the black hole, never decreases [Bek73, Bek74].

Second, almost concurrently with this suggestion, another proof was provided for certain laws of "black hole mechanics" [BCH73] which apply to stationary black holes and which are direct mathematical analogs of the zeroth and first laws of thermodynamics. In this proof the role of thermodynamic magnitudes such as energy E, temperature T, and entropy S, is played by magnitudes associated with the black hole, such as the mass M, a constant times the surface gravity κ and a constant times the area A, respectively.

The formal mathematical analogy, however, cannot be interpreted as carrying any physical significance in the classical arena, as the physical temperature of a black hole in that arena is absolute zero, and so there can be no physical relationship between T and κ. For this reason, it would also be inconsistent to endow a physical meaning to the relation between S and A. This caveat, however, is lifted when one considers quantum matter in curved spacetime. In 1974, Hawking [Haw75] made the surprising discovery that the physical temperature of a black hole is actually not absolute zero: as a result of quantum particle creation effects, a black hole radiates to infinity all species of particles with a perfect blackbody spectrum, at a temperature (in the natural units with $G = c = \hbar = k = 1$)

$$T = \frac{\kappa}{2\pi}. \tag{8.1}$$

Thus, in the quantum arena, $\kappa/2\pi$ truly is the physical temperature of a black hole, not merely a quantity playing a role mathematically analogous to temperature in the laws of black hole mechanics.

One might think that this entire line of thought which takes thermodynamics too seriously [Cal01] is just another wrongheaded attempt to save the second law "no matter what" (see Section 7.2.2). But regardless of what opinion one may have of Bekenstein's suggestion to assign thermodynamic entropy to the horizon area, or of the generalized second law that is associated with it, the black hole area law serves as a benchmark for all contenders to the solution of the quantum gravity problem as a feature that should be reproduced from their first principles.

The task is not easy, as it is still unclear what (and where) are the degrees of freedom responsible for the black hole entropy (see [Wal01], Section 6.2). After all, the metaphorical picture of the horizon's being divided into small "tiles" of a fixed size, with each tile carrying roughly one bit of information,[8] may be appealing to some (see Chapter 3 and [Bou02]), but from a strictly physical perspective one must give at least an indication as to what degrees of freedom carry this information.

Several suggestions have been put forward in this context. The first [BKLS86] is to focus on the correlations that exist between the degrees of freedom inside the black hole and outside its horizon. As a result of these correlations across the event horizon, the state of a quantum field when restricted to the exterior of the black hole is mixed, and one could use von Neumann's entropy $-\text{Tr}[\rho \ln \rho]$ to calculate

---

[8] In order to get the numerical coefficient right, the tile size would have to be around 10^{-65} cm^2, that is, it would have to be of order unity in the natural units [Sor98].

the thermodynamical entropy. However, in the absence of a short distance cutoff, the von Neumann entropy of any physically reasonable state would diverge. If one now inserts a short distance cutoff of the order of the Planck length, one obtains a von Neumann entropy of the order of the horizon area, A.

This additional motivation for a minimal length of the order of the Planck length might seem welcome to the quantum gravity phenomenologist, but she should not celebrate just yet, as there are many conceptual problems with this approach.

First, it is unclear why only the vicinity of the horizon is singled out as relevant to the black hole entropy. After all, according to orthodox quantum theory, entanglement does not decrease with distance, and so the same corrections are also expected far from the vicinity of the horizon. Second, and more serious, this approach relies on the premise that von Neumann's entropy is thermodynamic entropy. The only argument for this equivalence is von Neumann's thought experiment [vN32] but, as shown in [HS06], this argument only works for systems with infinite degrees of freedom and fails otherwise. The impasse here is clear: on the one hand one needs to remove the ultraviolet divergence of the von Neumann entropy by discretizing the possible modes of the state; on the other hand, this discretization prevents one from identifying the von Neumann entropy as thermodynamic entropy. Third, another serious dis-analogy between the case at hand and entropy in statistical mechanics and thermodynamics is that in the latter, ergodicity is often cited as a justification for the ability to count the microstates in a volume of phase space and to endow an equilibrium state with the highest entropy. But this model, while naturally justifiable within a classical framework, where the system really is in a definite (albeit unknown) microstate at any instant, is less justifiable in the quantum case, where, without decoherence, there is no reason to believe that the state would be in any eigensubspace of, say, the temperature operator, let alone evolve in accord with the desired Markovian dynamics.

A better suggestion is to locate the degrees of freedom responsible for the black hole entropy in the fundamental building blocks from which the horizon area (or its geometrical shape) emerge in theories of quantum gravity such as LQG or the causal set approach. The advantage of this suggestion is that it evades the impasse above by not just counting discrete quantum states of physically continuous variables, but by counting discrete physical elements. In the causal set approach, one might count the number of causal links crossing the horizon [Dou99]; in LQG one might count the number of loops cut by the horizon [Rov96, ABCK98]. In any case of this sort, the result would be something like the horizon area in units set by the fundamental (Planck) length. This result also meshes well with the metaphorical picture behind holography (see Section 7.2.2).

Finally, String theory gives, to date, the most quantitatively successful calculations for the Bekenstein–Hawking result (see, e.g., [SV96, Hor98]). What one counts in this case is the number of string states in the so called "weak coupling" limit of String theory [Car00], and yet, as is generally the case with String theory, it remains a challenge to understand in what sense the weak coupling states could be giving an accurate picture of the local physics occurring near (and within) the region classically described as a black hole [Wal01].

One may thus count (pun intended) the ability of approaches to quantum gravity such as LQG, the causal set, or String theory to reproduce the Bekenstein–Hawking result as evidence for their consistency. Note, however, that this proof is only indirectly related to the continuum structure via the mathematical analogy between "black hole mechanics" and thermodynamics, and, strictly speaking, in the classical general relativistic arena this analogy does not carry through. A more suitable description of this situation would therefore be to see the relation between black hole thermodynamics and quantum gravity as yet another example of the principle/constructive distinction (see Chapter 6), on a par, say, with the relation between thermodynamics and statistical mechanics.[9]

8.3.2 The 3 + 1 dimensions of classical spacetime

Under the weak version of the dynamical approach to spacetime we have presented in the previous chapters, certain features or aspects of spacetime geometry, for example, the symmetry group that characterizes it, or the notion of point-coincidence that underlies metrical relations, may depend on dynamical considerations. Another feature that is claimed to be dependent in this way is dimensionality.

The requirement that a background independent quantum gravity theory should possess the correct semiclassical limit, given by a macroscopically four-dimensional spacetime with microscopic quantum fluctuations, is non-trivial. In particular, the 3 + 1 dimensions of spacetime in the macroscopic domain have been demonstrated to emerge in computer simulations from the Causal Dynamical Triangulations approach to quantum gravity [AJL04], which is a framework for defining quantum gravity non-perturbatively as the continuum limit of a well-defined (regularized) sum over geometries [AJL10b]. It is hoped that in the mature theory these quantum geometries could be shown to emerge from an even more fundamental structure such as causal sets.

The fundamental length appears in this approach too, but contrary to the other approaches we have surveyed here, it vanishes in the limiting process when the individual discrete building blocks are literally "shrunk away" ([AJL10a], p. 188). This means that the fundamental length is merely regarded here as an intermediate regulator of geometry, as in the renormalization program (Section 4.4).

---

[9] In this respect, and as already hinted above, it is safe to speculate that many conceptual problems that saturate the foundations of statistical mechanics would reappear here. Philosophers of physics willing to delve into these intricacies would have their hands full.

Without entering into too many technicalities, for our purpose here it suffices to say that the way in which "spacetime as we know it" ([AJL10a], p. 191) emerges from the fundamental building blocks of this approach (the quantum geometries, summed along a path integral) is similar to the way certain macroscopic features of a physical system emerge in condensed matter physics from a lattice in the limit where the lattice's regulator vanishes. In the case under consideration here, one defines (and measures) geometric quantum observables, evaluates their expectation values on the ensemble of geometries, and draws conclusions about the behavior of the "quantum geometry," generated by computer simulations. The remarkable result of these simulations is that the macroscopic shape generated from this sum is that of the well-known 3 + 1 de Sitter universe [AGJL08] which we believe we inhabit.

8.3.3 The positive cosmological constant

Among the problems that modern cosmology faces is the existence of the so called "dark energy," which accounts for roughly 70% of the effective energy density of the universe, does not cluster like ordinary matter, and has negative pressure. A common, even popular, explanation for this existence is a positive cosmological constant Λ.

One of the problems with a positive cosmological constant, however, is that its calculated value in the natural units is some 122 orders of magnitude *larger* than its observed value (for this and for other issues concerning Λ, see, e.g., [RZ02, Ear03a]). Many thus believe that no theory could naturally predict this tiny value for Λ without predicting that it would vanish entirely.

A possible remedy to this problem comes from the causal set approach [ADGS04], where it is assumed that Λ is subject, just like the energy in ordinary quantum mechanics, to quantum fluctuations. This assumption, together with the randomization of the number of events per fundamental volume V, necessary for the consistency of this approach with local Lorentz invariance (see below), yields $\Delta\Lambda \approx 1/\sqrt{V}$. One then assumes, on pain of consistency with observed data, that Λ fluctuates around zero, and derives the result that Λ is (and always was) of the order of the critical energy density (in natural units). Note that in this approach one only answers the question why Λ is not exactly zero; the other question, namely why Λ is so nearly zero, is left open.

8.3.4 The (limited) power of consistency

These results demonstrate the unexpected advantages of making the assumption of fundamental length (here, fundamental area or volume) in solving several open questions in modern spacetime physics. And yet, critical readers will notice that there is quite a bit of fine-tuning in some of these solutions. Over and above the fact that geometry is assumed, as it were, all the way down (Chapter 7), when one knows what features of the geometry one would like to reproduce at the macroscopic scale, one can make suitable choices of parameters at the more fundamental level to arrive at the desired reproduction.

This is yet another reason, in addition to thesis *L*, why these results should be interpreted along the lines of the weak version of the dynamical approach, i.e., as mere consistency proofs, and not as strict derivations. These proofs are still important not only from the methodological perspective, but also because they carry a certain amount of epistemological weight. We now know, for example, what it means to say, from a microscopic standpoint, that a certain geometry satisfies Lorentz invariance: in the causal set approach, it means that, on average, there is one fundamental event per unit Planck volume or, in other words, that causally connected events are uniformly distributed in spacetime.

Let us suppose one could concoct similar proofs for all of the rival contenders to the solution of the quantum gravity problem. How, then, one may ask, are we to choose between them? For scientific progress, it seems, consistency proofs are necessary but not sufficient; something more is required to convince the scientific community to choose one putative building block (loops, causal sets, strings) over the others.

The requirement for new predictions, or deviations from the well-established principles of QFT, such as locality, relativistic causal order, unitarity, and Lorentz invariance at some scale, is important not only for reasons of theory choice; it will also allow quantum gravity phenomenologists to elevate the question of spatial discreteness from a purely metaphysical question to an empirical one. The problem, however, is that they must do so in such a way that these violations will either be confined to the fundamental length scale alone, or, if they do "trickle up," then they will somehow be deemed "rare," so that the model as a whole can still be consistent with the empirical evidence that led to the establishment of those principles in the first place. This, as we are about to see, turns out to be a very difficult task.

8.4 The perils of innovation

8.4.1 *Unus pro omnibus, omnes pro uno*

First let us remind ourselves what do these four principles, which, among others, characterize QFT, formally mean.

- *Locality* is a requirement that one imposes on interactions between two physical systems, namely, that they should occur at the same spacetime point. Formally, in QFT (or in any other field theory) this means that the interaction term in the Lagrangian of the theory contains only delta functions, functions of position at one (and only one) spatial point.

- *Relativistic causal order* is the requirement that a spatiotemporal event can influence another event if and only if the two events are time-like or null separated, and the first ("cause") is located inside (or on the surface of) the other's ("effect") past light cone, so that causal influence outside the light cone, sometimes referred to as "superluminal signaling," is not permitted. Once imposed, all observers should agree on this order and on the resulting no-signaling condition. In QFT this requirement is formally satisfied with the conditions of microcausality and cluster decomposition.

- *Unitarity* is a condition that in QFT ensures that probability (or the norm of the quantum state) is conserved through the dynamical evolution of the state. One can also view it as a symmetry condition, that ensures time reversal invariance of the dynamical evolution of the state. In QFT, this requirement is maintained by the unitarity of the *S*-Matrix, which conserves the probability of incoming and outgoing particles in a scattering process.

- *Local Lorentz invariance* is a relativistic symmetry condition, which ensures that experimental results are independent of the orientation or the boost velocity of the laboratory through space, so that observers in different reference frames related by Lorentz transformations will agree on these results in general (and on the constancy of the velocity of light in particular), thus preserving the principle of relativity. Agreement here means agreement that a specific pattern of point-coincidences occurs, and the measurement results in each frame are the Lorentz transform of the other.

Let us now inquire as to what are the possible consequences of introducing the notion of fundamental length into QFT with respect to each of these principles, comparing along the way the problems that were encountered in the attempts to introduce it into QED in the 1930s through the 1950s with those which arise today in current theories of quantum gravity.

8.4.2 *Locality and relativistic causal order*

As already noted by Dirac, a particle with a finite extension leads to breakdown of relativistic causal order at the length scale of its radius, as signals arriving at the particle's shell would travel instantaneously *inside* it (see Section 4.2.2). Attempts to introduce fundamental length into field theories in the 1930s have further shown that locality and relativistic causal order are violated at length scales of the order of the fundamental length (Chapter 4). In non-local field theories, for example

(Section 4.3.2), the Lagrangian explicitly contains interaction terms which depend on a form factor, namely, a weighted function of the position at two or more spatial points (see, e.g., [McM48]). The open question in the 1950s was whether it was possible to violate these two conditions "in the small," while maintaining them "in the large." Formally, this could be done, for example, if momentum space obeyed several continuity conditions [CP53]. In particular, large scale relativistic causal order was ensured if derivatives of arbitrary order existed of the form factor in momentum space. The problem, however, was that once gauge invariance and unitarity were imposed on the interactions, these continuity conditions made the self-energy diverge again [PU50, Pau53, CP54, SW54, Mar73], making the entire exercise moot.

In a way these results suggest that

... contrary to what one may have expected intuitively and contrary to the deceivingly lighthearted manner in which computational tricks as cutoffs and regulators are sometimes interpreted as basic physical concepts, causality is an extremely rugged property and that a notion of "a little bit non-local" or "a little bit a-causal" is not much more sensible than "a little bit pregnant."

([Sch05], pp. 94)

These lessons notwithstanding, the difficulties in unifying QFT with gravity have prompted many physicists to continue contemplating the abandonment of locality:

...It seems natural to suppose that the constant *l* [the fundamental length – AH] must appear when the third postulate, the postulate of locality, is abandoned. It is this assumption that is the basis of nonlocal field theory.

([Kir67], p. 695)

Of course the answer to the main question, as to how nature is actually constructed – in a local or in a nonlocal way – can be given only by nature itself. The deciding word in settling this question, the most important in the physics of elementary particles, belongs to experiment.

([Kir67], p. 699)

And more recently:

In all previous revolutions in physics, a formerly cherished concept has to be jettisoned. If we are poised before another conceptual shift, something else might have to go. Lorentz invariance perhaps? More likely, we will have to abandon strict locality.

([Zee03], p. 522)

In quantum gravity only a few models that could yield new phenomenological consequences have been proposed so far. Among these are models based on DSR, which violate locality [SU03, Hos10] by introducing another observer-independent constant of mass of the order of the Planck mass to STR. This introduction may result, via modified dispersion relations, in an energy or energy density dependent velocity of light. The crucial point here is that there seems to be no way to contain these violations and prevent them from "trickling up" to the macroscale, without abandoning the goal of distilling new predictions from the models.

Let us see why. For free, non-interacting systems, the DSR deformations leave the velocity of light intact and therefore there is no inconsistency with the conventional structure [Hos06]. Regardless of what one may think about the merit of free field models, either in standard QFT or in DSR, the important point for our purpose here is that one should not expect a model of this sort to yield any new predictions, and so it should be regarded, again, just as a consistency proof (we shall say more on this below, as this point is closely related to the principle of Lorentz invariance).

For interacting systems, the deformations lead to dependence of the velocity of light on either energy or energy density. As in the case of quantum optics discussed above (Section 8.2), the former model yields new predictions that could be tested *in practice* but, at least according to current data, also turns out to be false [Hos07]. In the latter model the effect is scaled down so that, while it may yield new predictions *in principle*, these cannot be tested *in practice*.

Furthermore, the inconsistency with locality may be removed by introducing the novel principle of "relative locality" (Section 7.5.1), as interactions still remain local in this approach even though a distant observer is only able to localize an event involving several particles with different momenta to within a region whose scale is proportional to the observers distance therefrom ([ACFKGS11], p. 10). But from the point of view of phenomenology, this move, again, may be regarded as a double edged sword: apart from making what is commonly considered an intrinsic property of spacetime dependent now on features of the geometry of momentum space, this solution has the dubious advantage of turning the apparent violations of locality into a non-physical coordinate artifact. The problem is that now it is also unclear what – under the new principle – remains of the prediction of energy-dependent velocity of light which was one of the major claims to fame of this approach.

Whether these suggestions could be developed further to yield other predictions that can be tested *in practice* is left to be seen. What is safe to conclude at this moment is that the situation nicely exemplifies the two delicate points I have been trying to draw attention to in this chapter. First, it is extremely hard to sacrifice certain well-established principles at one scale without violating them at all scales. Second, settling for *in principle* violations that, given their negligibility, may never be tested *in practice*, is of no avail if one wishes to turn the question of discrete space into a truly empirical one.

Another interesting way in which LQG may break locality is via the notion of "disordered locality" [MS07]. The idea here is that discrepancy may exist between the notions of locality at the microscale and at the macroscale. Locality at the level of the fundamental building blocks of the theory is defined, for example, by the connectivity of their combinatorial structure; locality at the macroscale is defined, for example, by the notion of the metric and by distance measurements. In some cases the two notions totally coincide, for example, when one associates to a classical metric g_{ab} semiclassical states with support on graphs which are embedded in space in such a way that only nodes that are within a few Planck distances of each other, as measured by the metric, are connected. In general, however, there is no reason to expect such an overlap.[10]

The possibility of this discrepancy between the two notions implies that there may be non-local effects present to some extent at the low energy limit of LQG, or of any other background independent theory with an underlying combinatorial structure such as the causal set approach. There are at least two ways in which the discrepancy could be manifest, depending on the state of the underlying spin network. First, in a single basis state, the emergent, coarse grained notion of a low energy metric could be insensitive to a small number of non-local links that connect "distant" (in the low energy sense of the word) nodes. Second, and more realistically, when the macroscopic quantum states involve superpositions of spin network states, there may be correlations which are local at the microscale but non-local in the metric that emerges at the macroscale. From what we know from non-relativistic quantum mechanics we expect that these correlations would be stable under the dynamical evolution of the state.

The notion of disordered locality raises the familiar two challenges we have encountered above. First, one would have to explain why these effects are rare enough not to disrupt local physics. Second, one should also indicate several (rare?) scenarios where these effects are manifest.

The second challenge is currently still unmet; what takes care of the first [PWS09] is the size of the fundamental degrees of freedom, and an assumption about their uniform distribution. Even in the presence of abundant non-local links, as many, say, as the number of baryons in the universe, the probability of their detection if they are randomly distributed is 10^{-120}. This small probability ensures that the effects are too small to detect, and will not differ from

---

[10] Apparently unknown to current quantum gravity theorists, a suggestion of this sort was already made by David Bohm who, since 1963 [Boh63], was interested in the notion of fundamental length, and believed the non-locality it harbors could explain what he called "the implicate order." In particular, in 1984, Bohm and Hiley [BH84] suggested constructing a continuous geometry on the basis of more fundamental and discrete Twistor and Clifford algebras. I thank Olival Freire Jr. for bringing this last paper to my attention.

thermal noise, even if we assume that the non-local links extend to terrestrial distances.[11]

Finally, the breakdown of relativistic causal order at the scale of the fundamental length has prompted at least three suggestions [Mea66, AM73, Ros84] to test it in the laboratory via the Mössbauer effect [Mos62], as the smearing of the light cone at the length scale of the fundamental length could lead to broadening of the spectrum of gamma rays (see Sections 2.3 and 5.5.2). The problem here is much more prosaic, as was already noted by Mead in the 1960s ([Mea66], p. 998). It would be extremely hard to isolate the mechanism behind the possible effect, and to pinpoint the fundamental length as its sole cause.

8.4.3 Unitarity

During the 1950s it became apparent that non-local field theories which introduce a Lorentz invariant form factor to the Lagrangian have a problem with unitarity [SW54], in the sense that once this principle is imposed on the S-Matrix, it destroys those conditions that tame violations of relativistic causal order at the macroscale. In other words, even if the form factor could be shown to be Lorentz invariant, one could not eliminate the divergences that plagued field theories without abandoning either one of these two principles.

In current theories of quantum gravity, unitarity enters into the discussion in a slightly different way. Here the problem arises in the context of the so called "black hole information loss paradox" (see, e.g., [Pre92, Ban95]). A piece of matter that enters into a black hole comes back as Hawking radiation (see Section 8.3.1), but this radiation, determined solely by the total mass, charge, and angular momentum of the black hole, is thermal, i.e., completely random, and so any information that the said piece of matter may have contained, is lost. Forever.

Lost, because if the information stays inside the black hole, eventually, as the black hole radiates and loses its mass, it would evaporate and would form a singularity. But this means that if the piece of matter started in a pure state, after being thrown into the black hole what remains is only the Hawking radiation which is always thermal, i.e., mixed. This transition, from a pure state to a mixed one, is non-unitary. We therefore arrive at a conflict between general relativity and quantum theory, unitarity being one of the fundamental principles of the latter.

There is a perfectly acceptable solution to this paradox within semiclassical quantum gravity. Think of a photon that escapes your room to the universe outside

---

[11] [PWS09] derive a contribution to the energy–momentum tensor from the presence of non-local links and the assumption on their distribution and arrive at a consistency proof for their density with the current expected order of dark energy. This proof is very weak, given the small probabilities we are dealing with, and the immensely slow rate of possible interactions of the non-local links with elementary particles. Furthermore, there is no way to test the model against alternative models.

it. While presumably in a pure state before the photon escaped, once it has gone, your room would be in a mixed state. Analogously, in a semiclassical analysis of the evaporation process, information loss can occur and is ascribable to the propagation of the quantum correlations into the singularity within the black hole, so that the final time slice fails to be a Cauchy surface for spacetime. Consequently, no violation of any of the local laws of quantum field theory occurs ([Wal01], pp. 29–30).

Here we encounter again the *a priori* negative attitude towards semiclassical quantum gravity that accompanied other arguments as the Eppely and Hanna thought experiment, as many in the quantum gravity community reject the semiclassical solution, and believe that only a *purely quantum* theory of gravity can resolve the paradox, and that it does so with holography [Bar09]. Holography preserves unitarity since the duality of the AdS/CFT ensures that the degrees of freedom that fall into the black hole are completely represented in the area of the horizon. Recently, however, doubt has been cast on this putative solution [AMPS12] and the debate that ensued has generated considerable hype, and even made it into the secular press [Ove13]. Others have suggested [SH09] that one can restore unitarity by eliminating the singularity. After all, such an elimination, as we have seen, is a consequence of almost all the scenarios of minimal length in approaches to the solution of the quantum gravity problem.

The problem, however, is not only with unitarity *per se*. Loss of unitarity has also been claimed to imply additional violations of other cherished principles, such as energy conservation [BSP84]. But also here the jury is still out: loss of unitarity implies a transformation from a pure state to a mixture, and this type of transformation is commonly described within quantum theory as "decoherence"; one way to mitigate the alleged violation of energy conservation is thus to demonstrate the possibility of decoherence without dissipation, or loss of energy to the environment.

Historically and conceptually, the idea of decoherence has its roots in the foundations of statistical mechanics and thermodynamics, where dissipation and energy exchange between the system and its environment are key concepts [Hag12]. Indeed, that decoherence (or dephasing, the delocalization of phase relations into the environment) involves dissipation is the presupposition of the (by now traditional) oscillator bath models within the approach of Leggett *et al.* [CL83, L$^+$87]. But in the more recent spin-bath models of decoherence, conceived within condensed matter theory, dissipation and dephasing are shown to be separate [PS00]. These new models of decoherence militate against the common view that ties the transition from pure to mixed states with energy loss because they involve non-Markovian dynamics. In the new models, the environment is entangled with the system so that it "tracks" its evolution with a kind of "memory." Spin-bath models therefore sever the traditional link between decoherence and energy dissipation, and therefore make the idea of non-unitarity appear less problematic [Unr12].

Since the information loss paradox starts from an assumption of the universality of unitarity, the discussion brings us closer to the search for violations of quantum theory, a topic familiar to philosophers of physics interested in the foundations of that theory, and commonly discussed in the non-relativistic context, and in particular in the context of the notorious quantum measurement problem (see Section 8.5). It remains to be seen whether evidence for loss of unitarity could be shown to be relevant to quantum gravity effects (most likely the context for testing these effects would be tests of (CP)T violations in natural kaons [EHNS84]) and, if so, how would quantum gravity phenomenologists react to lack of evidence, once the proposed experiments had been performed.

8.4.4 Lorentz invariance

STR has taught us that "Moving clocks run slowly; moving sticks shrink" [Mer05], Ch. 6). What happens, however, when a stick has the minimal Planck length, so it cannot shrink any further when set in motion relative to an observer? Put differently, does Lorentz invariance break down when spacetime becomes discrete? Note that one need not necessarily interpret the minimal length ontologically in order to appreciate the problem here, for it arises also when the minimal length is interpreted epistemically as a limitation on spatial resolution. In the latter case the question is whether Lorentz boosted observers would all agree on the value of the fundamental length as constraining the resolution of their spatial measurements.

The requirement that all observers agree on the constant value of the fundamental length may not be easy to satisfy but, as was (cryptically) argued already by Wataghin [Wat30b] in the 1930s, and shown later by Snyder [Sny47a] in the 1940s, it is not impossible. The question that remains open is whether current theories of quantum gravity can follow suit and incorporate spatial discreteness without breaking Lorentz invariance. This symmetry is so tightly constrained by current experiments to so many orders [Mat05], that many seem to ignore the fact it will never be tested at all orders, and prefer to think of it as an exact symmetry of nature ([Mat08], p. 15).[12]

In order to demonstrate the difficulties involved in this task, let us focus on a recent argument due to Rovelli and Speziale [RS03] that, in the context of LQG, aims to convince us that *in principle* (and not only *in practice*) the theory is

---

[12] In fact, it is probably safe to say that DSR motivated linear modifications of the dispersion relations are already ruled out by the data, but quadratic, or even higher order, modifications may not be so constrained (for opposing views on the matter, and on whether these modifications entail the notion of a preferred frame, see, e.g., [AC03b] and [CPS10]).

consistent with Lorentz invariance, at least to the extent that it remains silent with respect to its violation, giving as it does no concrete predictions thereof.

The putative compatibility of LQG with Lorentz invariance

The argument begins by noting the "simple minded" intuition ([RS03], p. 1): since length transforms continuously under a Lorentz transformation, a minimal length L_P (or a minimal area A_P, which is the corresponding measure in LQG) is going to be Lorentz contracted; if an observer at rest measures a minimal length L_P, a boosted observer will then observe the Lorentz contracted length $L' = \gamma^{-1} L_P$ which is shorter than L_P, and therefore L_P cannot be a *minimal* length (here $\gamma = 1/\sqrt{1 - v^2/c^2}$ is the Lorentz–FitzGerald contraction factor). Simple minded as it may be, the apparent conflict between discreteness of length and Lorentz invariance has had a large resonance on quantum gravity research. It appears, for example, as a motivation for DSR deformations of the Lorentz symmetry ([MS02, AC02c], or for the breaking thereof [GP99].

And yet according to Rovelli this intuition "is wrong, because it ignores quantum mechanics" ([Rov04], p. 316). The idea how to resolve the apparent conflict is the following. First one should note that in LQG length, area, and volume are not classical quantities; they are quantum observables. As a result, a boosted observer who measures these quantities will measure different observables. These boosted observables generally do not commute with the original observables that are measured by an observer at rest, i.e., if the system is in an eigenstate of L (the length operator at rest), generally it will not be in an eigenstate of L' (the boosted length operator). Therefore, there will a distribution of probabilities of observing different eigenvalues of L'. Put differently, the quantity of length will have an unsharp value for the boosted observer. Thus the mean value of the length can Lorentz contract, while the minimal non-zero value of the observable remains L_P.

The motivation for this alleged resolution comes from Snyder's famous counterexample to the above intuitive conflict between minimal length and Lorentz invariance, namely, spatial rotation symmetries of angular momentum in non-relativistic quantum mechanics [Sny47a]. This example at least demonstrates that discreteness of spectra is not sufficient to violate a continuous symmetry.

Consider a quantum system with total spin one. Suppose that an observer measures L_z and obtains the eigenvalue $L_z = \hbar$. Does this mean that a second observer, rotated by an angle α, will observe the eigenvalue $L' = \cos \alpha \hbar$? No. It is the mean value of L'_z which will be $\cos \alpha \hbar$, and not the eigenvalue. "States and mean values transform continuously with a rotation, while eigenvalues stay the same" ([Rov04], p. 317). What remains to be shown, and is shown in detail in [RS03], is that, as in the case of angular momentum, one can obtain unsharp values for the relevant boosted observable. This is done by proving that $[A, A'] \neq 0$, i.e., that the area observable at rest A and its Lorentz transform A' do not commute.

This result prevents LQG from making verifiable predictions about the continuous symmetry, in a similar way in which measuring the position of a particle in non-relativistic quantum mechanics prevents us from sharply predicting its momentum. It is therefore improper to say, conclude Rovelli and Speziale [RS03], and with them many in the quantum gravity community, that the symmetry breaks when acting on operators with discrete spectra. As we shall see below, however, this generalization from Snyder's example to LQG must be analyzed with great care.

Area measurements

In a couple of penetrating papers on the fate of Lorentz invariance in discrete spacetime [AC02a, AC03c], Amelino-Camelia presents a forceful objection to [RS03] which points at a dis-analogy between Snyder's example and the area measurement in LQG. In the former case, one can argue for compatibility between continuous symmetry and discrete spectra, but the reason for this compatibility is rather involved: it is only because a measurement of one component of the angular momentum, say L_x, introduces (in general) a significant uncertainty concerning the other components, L_y and L_z, that one can claim that quantum theory gives no verifiable predictions on the fate of the continuous symmetry ([AC03c], pp. 25–28). But one must identify at least one procedure that is suitable for a sharp measurement of L_x and that, at the same time, does not require sharp information on L_y and L_z, otherwise the reason behind the above undetectability remains ambiguous, as it could still be viewed as resulting from some contingent, practical inability to construct a suitable measurement that can reveal the fate of that symmetry, and not from some *in principle* features of the theory.

Let us look at this more closely. It can be shown ([AC03c], p. 27) that, in the case of angular momentum, one can still measure L_x without using any knowledge of L_y and L_z. However, this requirement seems to fail for the case of length (or area, or volume) measurements in LQG! The reason for this failure is the following. In order to achieve compatibility between Lorentz invariance and a minimal surface in LQG, the correct non-commutativity that must be demonstrated is between the area operator of the surface and the *velocity* of the surface, i.e., $[V, A] \neq 0$ ([AC03c], p. 30).[13] But what happens if a sharp measurement of A, which induces an unsharpness in the measurement of V, requires a sharp value of V in order to be performed in the first place?

---

[13] The Lorentz boost by a velocity V_0 leads to a transformation $\{V, A\} \to \{V', A'\} \equiv \{(V + V_0)/(1 + VV_0/c^2), A\sqrt{(c^2 - V'^2)/(c^2 - V^2)}\}$. The fact that this Lorentz boost transforms the area A into the area A' in a way that is continuous and *depends* on the velocity of the surface leads to the necessary condition $[V, A] \neq 0$.

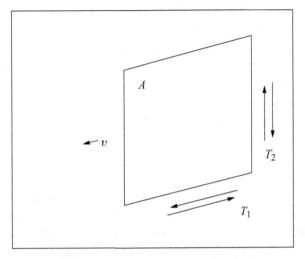

Figure 8.1 Area measurement with clocks.

Admittedly, there is no general argument from which one may conclude that all conceivable measurement procedures of the surface's area A are dependent on the knowledge of the surface's velocity V, but coming up with an appropriate independent measurement procedure seems to be a difficult task, as "all the commonly considered length measurement procedures do require sharp knowledge of the velocity of the ruler in order to achieve a sharp measurement of its length" ([AC03c], p. 30).

Here is an example of a measurement procedure that does require knowledge of the velocity ([AC02a], pp. 46–48). Imagine that we measure the area of a surface using the time-of-flight of two light bursts (assume for simplicity that we have previously established that the surface is rectangular, so that by measuring two sides one can obtain the area). The area of the surface should be obtained from two time-of-flight measurements T_1 and T_2. However, it is not sufficient to measure T_1 and T_2 to obtain an area measurement: it is also necessary to know the velocity of the surface! For if the surface is at rest the area will be deduced from the (T_1, T_2) measurement as $A = T_1 T_2 c^2/4$. But if the surface is moving with speed V along the direction of the T_1 measurement one would instead deduce from the (T_1, T_2) measurement that $A = T_1 T_2 (c^2 - V^2)/4$ (see Figure 8.1).[14]

One may try to object by saying that the formalism of QM need not be constrained by concrete models of measurement, and that it is sufficient to argue that an

---

[14] This assumes that the T_1 measurement is itself independent of the knowledge of the speed of the surface. In practice it is most natural to set the clocks in rigid motion with the surface, and then the T_1 readout would have to be corrected *in a V-dependent way* by the observer, since the relevant clock is not at rest with respect to the observer. The additional $\sqrt{1 - V^2/c^2}$ dependence does not change the nature of the argument, and one can therefore adopt the simplifying assumption that T_1 is measured in a V-independent way.

appropriate observable "exists," the idea being that any observable has physical significance, not only those for which we have concrete models of measurement. The point could be strengthened by the idea that for any observable, formally there is always some Hamiltonian that will couple the reading states of a measuring device to its eigenstates.

The problem, however, is that this assertion (that an area measurement procedure "exists" in LQG which is compatible with Lorentz invariance), does not strengthen the argument in [RS03], but rather weakens it; after all, this argument is supposed to depict the undetectability of the putative violations as a matter of theoretical *principle*, and not just as a contingent matter of *practice*, so it had better not depend on the latter. For why should one devise a clever argument to begin with, if all that is needed for the protection of the theory is that we do not know how to build a measurement apparatus that can detect the said violation? If a gun that can never hit a target is hidden so well in the sand on some island so remote, then why should I worry about the specific mechanism that ensures the gun is so unreliable?

So, in the case of [RS03], saying that there "exists" a Hamiltonian "somewhere" in the Platonic realm of Hamiltonians (without telling us what this Hamiltonian is), that can realize the measurement without knowledge of the velocity, can in no way serve as possible defense of an argument that is supposed to make violations of Lorentz invariance unobservable *in principle*. Faced with a non-constructive proof for the existence of the suitable Hamiltonian, one could just argue that these violations are simply too rare to be observed (this is, in fact, the strategy of the causal set approach; see below), or unlikely to be observed (because the construction of the Hamiltonian that detects them is "hard"), and leave it at that, thus making the argument moot, that prevents us from seeing these violations *in principle*.

No doubt there is a great temptation to prove that violations of some cherished physical symmetry can never be detected as a matter of *principle* in one's theory (although in this way one runs the risk of embarrassing one's theory if the symmetry turns out to be non-fundamental after all – a true empiricist never says never), but this proof had better not rely on some contingent matter of fact, for example, the practical inability to construct the apparatus that can detect such violations. And so, without a specific constructive proof for the measurement procedure, the argument in [RS03] remains incomplete.

A more plausible way to address the practical constraint of velocity dependence of area measurements is to accept it, but even this move still leaves the argument incomplete. Here is why. One can postulate that the discrete area spectrum applies only to areas at rest, i.e., sharply determined to have zero velocity ([RS03], pp. 7–9), so that only minimal *proper* areas are consistent with Lorentz invariance. This solution may work for a single, free, non-interacting observer who always measures her proper area, knowing her velocity to be $V = 0$. For more general scenarios that involve interactions or multi-particles, however, this solution is insufficient, since consistency with classical Lorentz symmetry dictates that an area eigenstate with well-specified velocity of the surface (in particular, "at rest," $V = 0$) must be an area eigenstate with well-specified velocity for all other inertial observers.

A simple analysis of the length measurement procedure described above can convince the reader that when two observers measure the length of a rod, they do so by using the same (light) clock. What is different, of course, is the analysis of the measurement readouts, which leads to different conclusions about the value of the distance/length being measured. Thus while both observers use the same measurement procedures in both the boosted and the unboosted cases to measure the same rod, they calibrate their clocks differently. On this view, it makes no sense to say that the same physical surface measured by both observers is an area eigenstate with a well-defined surface velocity only in the unboosted frame, for it would be tantamount to declaring two types of clocks, namely, one "at rest" and one "in motion." This is yet another dis-analogy with Snyder's counterexample on which [RS03] relies, where the measurement procedure of the angular momentum components involves measuring one and only one component of angular momentum. Instead, a surface measurement procedure which relies on the velocity of the surface involves measuring the equal-time area projection onto the two reference frames.[15]

I think it is safe to conclude that [RS03] may have demonstrated the consistency of LQG with Lorentz invariance for one observer who measures her own proper area (as Wataghin argued in 1930 [Wat30b]). But until they succeed in giving a faithful description of the operators of all observers in terms of the operators of this one observer (so that the picture of a world sheet measured by two observers above could be relaxed) or, alternatively, identify an area measurement procedure that does not rely on knowledge of the velocity, the jury is still out with respect to the status of this symmetry in LQG.

It goes without saying (but I shall say it anyway), that putative *in principle* violations of Lorentz invariance, such as, for example, the deformations of the kind suggested by DSR, need not entail observational consequences *in practice*. Here also, quantum gravity phenomenologists may be able to fine tune their models to regain consistency with the ever more tight constraints that tests of Lorentz invariance have been producing, and a methodological debate would then ensue (e.g., [CPS+04]) on the merit of such a strategy.

Other approaches to quantum gravity achieve consistency with Lorentz invariance in a more straightforward way. String theory has no problems doing so, at least in the so called "weak coupling" case, as its entities are constructed on the background of Minkowski spacetime [Hor05]. The causal set approach, as we have

---

[15] Think of a set-up similar to Bell's thread in Section 6.2.2 without acceleration, where B and C are drifting with constant equal velocities relative to A, and imagine that A and B are both trying to measure the distance between B and C.

seen, introduces an assumption of uniform (random) distribution of causal events in a unit spacetime volume (actually, in this approach this is how spacetime volume is *defined*) and, via coarse graining [DHS04], it ensures that macroscopic violations of Lorentz invariance would be rare enough and practically unobservable, in accord with current constraints.

8.4.5 Other ideas

Taking stock, among the current approaches to quantum gravity, it is hard to find someone who would risk her neck, as it were, and seriously explore the breakdown of one or more of the above four principles we have been discussing. Rather, efforts have been focused so far on constructing consistency proofs for their validity.

To sum this up in one (long) sentence, breakdown of unitarity may be blocked by holography, nobody would seriously contemplate breaking locality or relativistic causal order by introducing an unobservable preferred frame, as this is considered the ultimate borderline between physics and pseudoscience, and, as we have seen, it is extremely hard to deform Lorentz invariance without breaking the former two principles.

Currently, it is only within the DSR approach that one finds proposals for deforming the Lorentz transformations that may lead to observable effects, thus elevating the model from a mere consistency proof to one that may yield novel predictions, but even in that approach it is still unclear whether such deformations carry any practical phenomenological consequences in a manner that does not already render the model false. Many thus believe that models predicting signatures of spatial discreteness should be constructed that respect Lorentz invariance and that do not break, at least not explicitly, the other three principles.

Two suggestions are worth mentioning here. The first [BS08] takes its cue from symmetries in crystals, as seen from two resolution levels, the micro and the macro, and an analogy that may exist between this situation and the relation between the fundamental granularity of spacetime and its smoothness at the macroscale. The idea here is that if the fundamental crystalline (granular) structure respects the crystal's macrosymmetries, no signature of the said granularity could be detected by looking at possible deviations therefrom. What we should look for, instead, are deformations in the crystal's macrostructure in situations where the crystal's global form is not compatible with the granular structure, as in these cases the surface will necessarily include some roughness, and thus, a manifestation of the granular structure will occur through the breakdown of the exact microsymmetry ([BS08], p. 3).

Translating from the crystal case to spacetime, if the latter fails to be exactly Minkowskian in an extended region, the underlying discrete structure could become manifest, affecting the propagation of the various matter fields. This situation is known to be described by a non-vanishing Riemann tensor (i.e., a finite curvature to spacetime) and, in contrast to the breaking of Lorentz invariance, a model that relies on this idea predicts no deviations from known physics in flat spacetime [CPS10]. This intuition motivates a phenomenological model that could be tested *in principle* which resembles variants of the Standard Model Extension [Kos04] but, in contrast, does not introduce a preferred frame.

The second idea, due to Sabine Hossenfelder [Hos12], is to circumvent the aforementioned problems of the deformations of the Lorentz transformations, and to postulate that the metric of the (flat) spacetime described by DSR is a superposition of terms which contain different invariant velocities of light.[16] This suggestion leads to an elegant structure, free of the problems that previous DSR models face, but it singles out the quantum measurement process as having several special roles:

We will assume that the process of measurement produces an observable that henceforth transforms under the c^*-representations [the extended Lorentz transformations with the additional parameter – AH]. The process of measurement also picks out one particular rest-frame that plays the role of a preferred frame once the measurement has been made.

([Hos12], p. 4)

The preferred frame is assumed to be the rest frame of the bath which environmentally induces the apparent collapse via decoherence. Apart from singling out this frame, and from carrying out "the transition from quantum to classical," the measurement also reduces the extended Lorentz symmetry (in the superposition) to the usual one; quite a number of roles for a single quantum measurement!

The model has the advantage of sweeping under the rug the problem with locality encountered above with previous suggestions to deform the Lorentz transformations, since after a measurement has been made in one frame, all observers would agree on the point-coincidence of events. It does, however, include the possibility of superluminal signaling and hence allows backwards causation (a normal Lorentz transformation can turn a superluminal curve into one going backwards in time). To evade the standard paradoxes associated with closed time-like curves, one can simply endow the measurement with yet another role, namely, that of picking out a preferred foliation, hence a direction in time.

There may be a lot to like in this proposal, but I can imagine philosophers of physics who are versed in the foundations of non-relativistic QM already moving nervously in their chairs, thinking probably that this "solution" just trades one problem for a list of many others. One has to be rather desperate to expect so much from the quantum measurement process, a process which for almost four

---

[16] Here the different c depend on a free parameter. The specific role of energy or momentum in this dependence is left open for future developments of the model.

generations has resisted a precise physical underpinning. In particular, just saying that "decoherence via interactions with the environment" takes care of all the problems will certainly not do.

It is by now common knowledge, at least among philosophers of physics [Bub97], and also among some physicists [Adl03], that decoherence – while describing formally how a pure state can turn into a(n improper) mixture – is an incomplete account of a transition of this sort, and must be accompanied by interpretations of quantum theory which analyze dynamically the measurement process [Mau01]. In addition, maybe less appreciated, but no less true, is the fact that decoherence (or "the transition from the quantum to the classical") does not pick a direction in time, since, given the unitarity of quantum dynamics, this asymmetry resides solely in the initial conditions of the universe [Hag12]. If these were such that branches of the said superposition of the different velocities of light would recohere one day, then so much for the preferred direction picked up by the apparent collapse.

A more pressing problem with this proposal is that it seems to militate against the possibility of ever deciding experimentally the question of spatial discreteness. Agreed, it may be premature to draw any phenomenological conclusions from the model proposed here but, at least according to its author, these are confined, again, to checking whether the free parameters of the model agree with current constraints laid by the experimental evidence for QFT. Remarkably, almost three generations later, Heisenberg's speculation about the inability to falsify practically the *in principle* consequences of minimal length (Section 5.4.2) has found its ultimate incarnation in the idea that these consequences are hidden in a quantum superposition of the speed of light.

8.5 There and back again

We have come a long way, on a bumpy ride, that took us on a journey through almost a century of attempts to introduce fundamental length into field theories. I would like to end this chapter, as well as this book, with an observation that might surprise those physicists who have been working on quantum gravity phenomenology, but might seem quite natural to philosophers of physics.

It is a little ironic that in the attempts to unify quantum theory with the general theory of relativity, one ends up with conceptual problems such as non-locality, superluminal signaling, and a preferred foliation of spacetime. All these problems are part of the ongoing debate on the "peaceful coexistence" of quantum mechanics with STR, challenged by the different solutions to the quantum measurement problem (see, e.g., [Alb92, Mau94]). This problem plagues all quantum theories, the non-relativistic as well as the relativistic theories [Bel90], but physicists were able to ignore it and make progress when all they had to do was to extract verifiable predictions from the quantum formalism, using it as an algorithm for the purpose of calculating probabilities.

While I certainly do not think that making progress in solving the quantum gravity problem requires a solution of the measurement problem, an attention to the methodological and the philosophical issues that surround the attempts to give a dynamical analysis to the measurement process may prove instructive. The quest for quantum gravity phenomenology bears, for example, many similarities, on different levels, to a research program known as the spontaneous collapse theories in which one attempts to construct an alternative to non-relativistic quantum theory that could physically describe the collapse process as intrinsic to a closed system [GRW86]. In these attempts one is trying to alter Schrödinger's equation, hoping to find minute deviations from the Born rule that could be tested in rare situations in the laboratory.

The similarity here is striking. First, the same problems with three of the aforementioned principles (unitarity is forfeited from the start), as well as with energy and momentum conservation, also arise here [Ghi11]. Second, at a strictly methodological level, quantum gravity phenomenologists and those theoreticians trying to construct phenomenological collapse models [Pea86, Dio89, BIA05, ABI05, AB09, Sta12] face the same challenges: both are looking for new, testable, predictions which are tightly constrained by the well-established principles of quantum theory and STR, both are being accused of fine-tuning the parameters of their models to evade conflict with current available data, and both are looking for rare experimental set-ups which are also extremely hard to control. Third, both research programs carry a similar philosophical significance. If successful, they may turn what was supposed to be a strictly metaphysical question into an empirical one. If small deviations from the Born rule could be found, thus verifying collapse theories, one could argue that the fundamental dynamics in nature is indeterministic and time reversal is non-invariant.[17] Similarly, if the quest for quantum gravity phenomenology bore fruit, then given the current theoretical landscape, it would give *empirical* support to the hypothesis of minimal length, which so far has been deemed a metaphysical speculation.

Speaking of speculations, there are those who have been claiming over more than a generation that the two research programs are actually related from the physical perspective, by invoking gravity as a physical mechanism for wave function collapse [Pen00]. Regardless of what one may think about this idea, one must admit that it has yielded an astonishing research program with some amazing experimental

---

[17] Recall that Schrödinger's equation is deterministic and unitary, and that the collapse process in standard quantum theory is not a fundamental physical process, described as it is only via the interaction of the quantum system with its environment. For a closed system such as our universe, therefore, there is no collapse.

set-ups [MSPB03]. But one need not adhere to this suggestion in order to appreciate the previous methodological and philosophical points.

The upshot, if there is one, is that apart from looking at the foundations of mathematics for hints and clues how to rebut *a priori* arguments against spatial discreteness (Chapters 2 and 3), quantum gravity phenomenologists may also like to consider the vast literature on the philosophy of non-relativistic QM. They might find that not only the mathematicians, but also the philosophers, have been there before, all along.

And what, you may ask, happened to the puzzle we started our journey with? Can we finally argue that the question of spatial discreteness is an empirical question?

Here, I am reluctant to say, the answer depends on the benchmark for these questions. If consistency proofs are sufficient to make a hypothesis qualify as empirical, then, by showing the hypothesis of minimal length to be consistent with field theories and their description of space as a continuum, one has turned metaphysics into physics, at least to the extent that spatial discreteness could now be claimed to be on a par with the continuum as a viable, alternative, description of the structure of spacetime. If, like me, one requires more than mere consistency proofs, since it seems that only new predictions will be able to decide the issue and to convince supporters of the continuum to abandon their view, then, I am afraid, we still have a long way to go.

9

Coda

Aleph

The struggle with infinity – both in space and in time – has always baffled mankind. In physics, the issue, in its most simple manifestation, is the difficulty in coming to terms with the notion of a dimensionless point, and with the divergent physical magnitudes associated with it. This struggle has also left traces outside physics. In one of Jorge Luis Borges' short stories, *The Aleph*, the protagonist, a fictionalized version of the author, comes across a point in space that contains all other points. Told by his nemesis in the story that anyone who gazes into it can see everything in the universe from every angle simultaneously, without distortion, overlapping or confusion, Borges is struck by what he calls "the hopelessness of the writer," and yet in the next few lines he unleashes one of his longest sentences ever, over two pages long; quite remarkable for a writer known for his economy of expression.

On this metaphorical note, the debate on the notion of fundamental length can be seen as serving as an *Aleph* point in the "universe" of the history and philosophy of modern physics; a point where most of the unresolved problems and the conceptual difficulties that we face today in our attempts to describe nature at its most fundamental level, converge. Unlike Borges, however, I shall keep my prose short even after gazing at this point, and in what follows I shall offer a short Q&A session that summarizes the paths taken in the study presented here.

Questions and answers

Q1. Why would one introduce the notion of fundamental length into modern physics?

Many are the motivations for doing so, not the least of which are philosophical in character.

From the purely metaphysical perspective, one may endorse a fundamental view of the world that requires that there are only finite degrees of freedom in a finite volume. This view is consistent with the so called digital metaphysics, popular among computer scientists and information theorists, and physically amounts to an upper bound on the amount of energy that can be localized in a given spacetime volume.

Such a limit on physical resources may also follow from an epistemological reasoning: according to traditional views such as operationalism and positivism, only practically measurable magnitudes, whose resolution is always finite, are physical. Infinite precision is deplored from this perspective as an unphysical myth, and the continuum is seen as merely a convenient abstract apparatus without any real correspondence in nature. That field theories, whose mathematical structure is a continuum, have so far been successful in describing nature is, for the finitist, only a temporary situation, which rests on the huge gap that exists between our current resolution capabilities and the most fundamental possible resolution.

These motivations are manifest in physics in the attempts to eliminate singularities and tame the divergences that have plagued field theories, both classical and quantum. The notion of fundamental length is thus the embodiment of finitist metaphysics and operationalist epistemology.

Finally, from a methodological perspective, one may pursue a unifying strategy that requires, on pain of methodological consistency, that gravity is quantized just as any other interaction. And *if* gravity is quantized, then one ends up with a limit on the ability to localize a particle to a radius below the Planck length. Note that the converse does not follow, namely, this limit on spatial resolution does not entail that the gravitational field must be quantized. The additional, independent, assumption required to complete this deduction is that quantum theory is indeed universal.

Q2. How is the notion of fundamental length introduced into modern physics?

Broadly speaking, one can eliminate singularities and remove divergences either dynamically or kinematically. The first strategy requires imposing an upper bound on momentum transfer in collisions between particles. Quantum mechanically, an upper bound on momentum means a lower bound on position. A dynamical cutoff can also be realized by introducing a curved geometry to momentum space. The second strategy imposes a finite measurement resolution on physical quantities. The two strategies become equivalent when one analyzes the measurement process dynamically as an interaction between two subsystems of the universe. The result is a characteristic length that each test particle carries, proportional to its own mass.

Around the 1930s the two strategies were commonly used in the attempts to eliminate the divergences that threatened the consistency of QED, notably by Bohr and Rosenfeld, Heisenberg, Born, March, and Wataghin. Adding gravity into the mix, the characteristic length also became minimal, since – as noted by Bronstein, Wheeler, Klein, Deser, and Mead, to name a few – no force other than gravity couples to anything material regardless of the mass involved. These considerations may lead one to impose an absolute uncertainty on spatial resolution, which is formally equivalent to letting two position observables become non-commuting.

The arguments for the limit on spatial measurement resolution depend on thought experiments that in turn involve some version of Heisenberg's microscope, or an extension of Bohr and Rosenfeld's argument for the measurability of field quantities to include gravity. Consequently, they all rely on the so called semiclassical "disturbance" view of the uncertainty principle. The "disturbance" view is widely discredited today among philosophers because it is believed to lend support to an epistemic approach to QM, according to which a quantum measurement reveals the values of existing, unknown variables. The whole point of interference phenomena and the confirmed violations of Bell's inequalities, or so the story goes, is that there are no such local hidden variables. Also widely criticized is the idea that the "disturbance" view implies that classical fields must be quantized, otherwise the uncertainty principle could be violated.

We have suggested a way to mitigate this criticism. First, even if one retains the epistemic view of quantum probabilities, one can remain agnostic about the metaphysics of the quantum state by taking measurement outcomes to be primitive events on a non-Boolean probability space. An example is quantum Bayesianism, according to which the measurement process is regarded as primitive, and quantum probabilities are interpreted as (non-Boolean) epistemic degrees of belief.

Such agnosticism transfers to an objective finitist view of quantum probabilities as (non-Boolean) physical transition probabilities. On this view probability signifies "how hard" it is to realize one state from another given a fixed amount of physical resources (energy, space, time), and the question of hidden variables does not arise since the difference between quantum and classical probabilities is seen as a difference in measure, rather than a difference in metaphysics. The upshot is that unless one insists on adhering to a specific metaphysical interpretation of quantum theory, there are no good reasons not to replace the "disturbance" view with an agnostic one (or with a more radical view that seeks to underpin non-commutativity physically by showing how it arises from inherent limitations on spatial resolution), as long as the interpretation of quantum probabilities remains an open question, and as long as the different approaches that answer this question succeed in reproducing the Born rule for all experiments done so far.

As for the necessity to quantize the gravitational field that allegedly follows from the thought experiments that establish the limitation on spatial resolution, we have suggested that the arguments for fundamental length in the period from the 1930s through the 1950s, which were motivated by the said thought experiments, could also be construed as arguments for the epistemic coherence of any putative theory that purports to describe phenomena that involve both QFT and gravitational effects. If this theory existed, it would have to be constrained by the additional physical restrictions on measurement resolution imposed by gravity (see Q1). In other words, the arguments for fundamental length could also be regarded as arguments for the limit of applicability of such a putative theory.

Whether proponents of these arguments were conscious of the problems of the "disturbance" view or of our alternative to it is, of course, doubtful, but at least some of them were clearly conscious of the lack of logical necessity for quantizing the gravitational field, and considered their thought experiments as supporting no such thing, interpreting them as they did along the lines we have suggested here, i.e., as supporting the epistemic coherence of any putative theory of quantum gravity, so that within its limits of applicability it would not preclude its own verification.

Q3. What role does fundamental length play in current approaches to quantum gravity?

In current approaches to quantum gravity, the role of fundamental length is, as before, the elimination of singularities and divergences; it enters, however, in a slightly different way.

- In String theory, the fundamental length is the characteristic string length (a function of its tension), and yet, since strings do not "live" in ordinary three-dimensional space, it is unclear how to translate this discreteness into physical (in contrast to purely geometric) spatial discreteness.
- In LQG, the fundamental length arises in the discrete spectra of observables that are singled out as representing geometrical notions of area and volume. In other words, spatial discreteness is a prediction of the theory, and not an assumption.
- In the causal set approach, the fundamental building blocks are spatiotemporal events whose causal structure, together with a combinatorial assumption that allows one to associate their number with the geometrical notion of "volume," yield the classical Riemannian metric. Here discreteness means "one event per unit Planck volume."
- In the Emergent Gravity approach there is no clear meaning to spatial discreteness, although in many of the condensed matter models on which this approach is based, the underlying structure is a lattice.

- Kinematical approaches, for example, DSR and non-commutative geometries, adopt some of the earlier strategies from the 1930s, such as absolute uncertainty and the idea (known as Born's Reciprocity) that the (now believed to be curved) momentum space has an upper bound. It is still unclear how these approaches relate to the dynamical approaches mentioned above.

Q4. How should we interpret the high momentum cutoff or the limitation on measurement resolution?

The short answer is that this interpretation remains a matter of taste, so long as there are no phenomenological deviations from the well-established principles that underlie field theories which (so far) perfectly describe spacetime as a continuum.

That a dynamical cutoff of high momentum does not entail a lattice structure to spacetime can be appreciated from two perspectives. First, from a philosophical perspective, the inability to resolve short distances is an epistemic limitation, and therefore cannot dictate the actual state of affairs. In other words, nature can be fundamentally continuous regardless of our finite measurement resolution. Second, from a theoretical physics perspective, a curved momentum space with a cutoff on high momenta does not translate easily into position space in any straightforward way (see Q5). This is probably the reason why many physicists, even those who proposed the said cutoff in the 1930s, chose to remain agnostic about the actual structure of space, and interpreted the cutoff as a limit on the theory's applicability, above which (in momentum space) or below which (in position space) additional physics may take place.

Q5. What problems does the notion of fundamental length generate?

Intuitively, and as was already acknowledged by Heisenberg in 1930, just naively introducing a lattice structure into three-dimensional space, or an absolute cutoff on momentum transfer in collisions, results in inconsistencies with continuous symmetries, conservation laws, unitarity, locality, and relativistic causality.

While partial consistency with Lorentz invariance and unitarity is established for free fields in the four-dimensional arena, conflict with causality and locality seems unavoidable. For this reason, modern physics, via the renormalization program, has chosen to evade these problems by taking the middle way between finitism and actual infinity (see below, Q7). In QFT the fundamental length is only characteristic, and not minimal, designating as it does the domain of applicability of the effective field theory at hand, without any commitment to a fundamental, or universal, cutoff.

In quantum gravity these problems persist, but one can also see the situation as an opportunity, and not as a reason for despair. In order to make the question of spatial discreteness an empirical one, it is at least clear now that one must let go of some of the characteristic principles of either quantum theory or STR. The problem is, as always, a problem of what to hold, and of what to let go.

Q6. What are the possible phenomenological consequences of fundamental length?

Breakdown of any of the principles that underlie QFT (e.g., locality, relativistic causality, unitarity, and Lorentz invariance) would have dramatic theoretical consequences, but the harder problem is to come up with phenomenological models that include fundamental length that could yield *in principle* new predictions, and would still abide by these principles at the scales at which these have been verified.

The issue at stake is thus not whether one or more of the above principles are violated; the issue at stake is how to violate these principles at one scale, presumably the Planck scale, while keeping the model consistent with the empirical evidence that supports their validity down to scales of 10^{-16} cm.

It is therefore not surprising that most of the current phenomenological models in quantum gravity are aiming at consistency proofs of the underlying discrete structure with the lower energy, continuous, description, and only very few of them dare to yield new predictions.

Given the robustness of locality, causality, unitarity, and Lorentz invariance, the best strategy that has emerged so far in this context is a combinatorial one. This strategy is common in theoretical physics in situations where one is trying to explain why certain possible phenomena, predicted by one's model, nevertheless remain unobserved.

Examples of this strategy are the explanation of the consistency of time reversible invariant statistical mechanics with the thermodynamic arrow in time, or the explanation of the consistency of the linear Schrödinger equation with the lack of entanglement in position basis for macroscopic objects. In these explanations, if one could call them so, one assumes a certain probability distribution on the event space, such that any deviation from the well-established results is deemed "rare." The advantages here are clear, as one can still predict new results in certain "rare" situations and, at the same time, declare an "on average" consistency with the verified principles one is actually holding as non-fundamental.

The open philosophical question here is whether this cautious combinatorial strategy could nevertheless turn the debate on spatial discreteness from a purely metaphysical one, beyond mankind's experimental reach, into an empirical one that can be decided by experience.

Q7. What philosophical lessons can we learn from the history of fundamental length?

First and foremost, on a meta-philosophical level, we have seen that many argument templates, for example, arguments that aim to demonstrate the consistency or the applicability of only one geometrical structure, be it the continuous or the discrete, find their way from the philosophy of mathematics into theoretical physics. We have also seen that arguments for (and solutions for problems with) fundamental length that were discussed in the context of QED and QFT reappear almost verbatim in current approaches to quantum gravity.

Moving down to the philosophical level, but still within the broad context of the history and philosophy of science, we learn that there are rare cases in the history of science, and the history of the quest for fundamental length is one of them, in which certain philosophical theses can be directly mapped onto theoretical practices.

This mapping is evident in the construction of field theories and the status of singularities therein, especially in QFT.

- Finitism in mathematics abhors singularities, and aims to eliminate them with limitations on spatial resolution.
- Platonism is quite happy with actual infinity and the continuum, not only as abstractions, but as real physical possibilities.
- The third way, the way that turned out to be the one chosen, supports potential infinity, and is a compromise between the above two extremes: while non-committal as to the ultimate, final, structure, it still fixes a certain length scale as if it were fundamental, and eliminates singularities *for all practical purposes* at that specific scale.

A more specific philosophical moral has to do with the limits of reduction, in our case the reduction of spacetime geometry to a more fundamental, dynamical theory, whose building blocks are discrete. In the attempts to construct a quantum theory of gravity, certain features of spacetime, namely, the symmetry groups that characterize it, its continuous structure, or the answers one may give to questions like "do two events happen at the same spacetime point?," may become dependent on the more fundamental theory, its building blocks, and the dynamics thereof.

As startling as they may seem, these result do not amount to the more radical claim that "spacetime disappears" at the Planck scale. As argued here (we have called this "thesis L"), quantum gravity theories must, and also do, as a matter of fact, single out certain theoretical magnitudes and designate them as geometrical or spatiotemporal objects, in order to make contact with the experiments that can verify their models. And since this designation is done by stipulation, nothing in the fundamental theory, its basic building blocks, or the dynamics thereof, fixes that reference *ab initio*. Geometry and spatiotemporal notions thus remain primitive even in quantum gravity.

For this reason, the achievements of current approaches to quantum gravity, impressive as they may be, should be interpreted along the lines of a weaker version of reduction, a non-eliminative one, wherein one constructs consistency proofs for the (dynamical) reducing theory with the (geometrical) reduced one. Apart from discovering certain features and properties of spacetime, these consistency proofs have the merit of explaining these features in the language and terms of the dynamical theories. For example, that spacetime is macroscopically (on average) locally Lorentz invariant means, under the causal set approach, that the building blocks of the theory are uniformly distributed in spacetime.

Another philosophical lesson is that the old methodological debate from the 1950s on the necessity of quantizing the gravitational field is still open. Contrary to quantum gravity folklore, and even with the notion of fundamental length, there is no logical necessity to quantize gravity (see Q1 and Q2). Semiclassical theories are physically possible, and the reason they are rejected is simply practical. Rather than arguing for such a quantization on an *a priori* basis, one should look for empirical signatures that could motivate it, keeping in mind that, at least relative to the energy scales we currently have access to, these empirical signatures may remain beyond our reach.

We can finally formulate our closing statement: a thousand years from now, when the rigid adherence to continuum geometry will have been loosened by centuries of discrete math, algebra, information theory, and computer science, the resistance to the notion of fundamental length will probably look like just another case of human myopia. Until then, not only will it be difficult to persuade the skeptic (one hopefully without any personal stake in the matter) that spacetime is not continuous, it will also be difficult to persuade her that the question is actually an empirical one.

Difficult, but, as argued here to an extended length, not impossible.

References

[A+82]   A. Aspect *et al.* Experimental Realization of Einstein–Podolsky–Rosen–Bohm *Gedankenexperiment*: A New Violation of Bell's Inequalities. *Physical Review Letters*, **49**:91–94, 1982.

[AAV93]   Y. Aharonov, J. Anandan, and L. Vaidman. Meaning of the Wave Function. *Physical Review A*, **47**:4616–4626, 1993.

[AB61]   Y. Aharonov and D. Bohm. Time in the Quantum Theory and the Uncertainty Relation for Time and Energy. *Physical Review*, **122**:1649–1658, 1961.

[AB09]   S. Adler and A. Bassi. Is Quantum Theory Exact? *Science*, **325**(5938):275–276, 2009.

[ABCK98] A. Ashtekar, J. Baez, A. Corichi, and K. Krasnov. Quantum Geometry and Black Hole Entropy. *Physical Review Letters*, **80**:904–907, 1998.

[ABI05]   S. Adler, A. Bassi, and E. Ippoliti. Towards Quantum Superpositions of a Mirror: An Exact Open Systems Analysis – Calculational Details. *Journal of Physics A*, **38**:2715, 2005.

[Abr04]   M. Abraham. Die Grundhypothesen der Elektronentheorie. *Physikalische Zeitschrift*, **5**:576–578, 1904.

[AC75]   T. Appelquist and J. Carazzone. Infrared Singularities and Massive Fields. *Physical Review D*, **11**:2856–2861, 1975.

[AC01]   G. Amelino-Camelia. Testable Scenario for Relativity with Minimum Length. *Physics Letters B*, **510**:255, 2001.

[AC02a]   G. Amelino-Camelia. On the Fate of Lorentz Symmetry in Loop Quantum Gravity and Noncommutative Spacetimes. *Preprint*, 2002. arXiv:gr-qc/0205125v1.

[AC02b]   G. Amelino-Camelia. Quantum Gravity Phenomenology – Status and Prospects. *Modern Physics Letters A*, **17**:899–922, 2002.

[AC02c]   G. Amelino-Camelia. Relativity in Spacetimes with Short-Distance Structure Governed by an Observer-Independent (Planckian) Length Scale. *International Journal of Modern Physics D*, **11**:35–60, 2002.

[AC03a]   G. Amelino-Camelia. Kinematical Solution of the UHE-Cosmic-Ray Puzzle without a Preferred Class of Inertial Observers. *International Journal of Modern Physics D*, **12**:1211–1226, 2003.

[AC03b]   G. Amelino-Camelia. Proposal of a Second Generation of Quantum-Gravity-Motivated Lorentz-Symmetry Tests: Sensitivity to Effects Suppressed Quadratically by the Planck Scale. *International Journal of Modern Physics D*, **12**:1633–1640, 2003.

[AC03c] G. Amelino-Camelia. The Three Perspectives on the Quantum–Gravity Problem and Their Implications for the Fate of Lorentz Symmetry. *Preprint*, 2003. arXiv:gr-qc/0309054v1.

[AC06] G. Amelino-Camelia. Anything Beyond Special Relativity? *Springer's Lecture Notes in Physics*, **702**:227–278, 2006.

[AC07a] G. Amelino-Camelia. A Perspective on Quantum Gravity Phenomenology. Presented at the PI Workshop on Experimental Search for Quantum Gravity, pirsa:07110057, min. 26:23, 2007.

[AC07b] G. Amelino-Camelia. Relativity, Still Special. *Nature*, **450**:801–802, 2007.

[AC12a] G. Amelino-Camelia. Born's Prophecy Leaves No Space for Quantum Gravity. *Preprint*, 2012. arXiv:1205.1636.

[AC12b] G. Amelino-Camelia. No Theory is Too Special to Question. *Nature*, **483**:125, 2012.

[ACFKGS11] G. Amelino-Camelia, L. Freidel, J. Kowalski-Glikman, and L. Smolin. The Principle of Relative Locality. *Preprint*, 2011. arXiv:1101.0931.

[ACS09] G. Amelino-Camelia and L. Smolin. Prospects for Constraining Quantum Gravity Dispersion with Near Term Observations. *Preprint*, 2009. arXiv:0906.3731.

[ACV89] D. Amati, M. Ciafaloni, and G. Veneziano. Can Spacetime be Probed Below the String Size? *Physics Letters B*, **216**(1–2):41–47, 1989.

[ADGS04] M. Ahmed, S. Dodelson, P. Greene, and R. Sorkin. Everpresent Lambda. *Physical Review D*, **69**(10):103523, 2004.

[Adl03] S. Adler. Why Decoherence Has Not Solved the Measurement Problem: a Response to P.W. Anderson. *Studies in the History and Philosophy of Modern Physics*, **34**(1):135–142, 2003.

[AGJL08] J. Ambjørn, A. Görlich, J. Jurkiewicz, and R. Loll. Planckian Birth of a Quantum de Sitter Universe. *Physical Review Letters*, **100**:091304, 2008.

[AI30] V. Ambartsumian and D. Ivanenko. To the Question of Avoidance of the Infinite Self-Counteraction of Electrons. *Zeitschrift für Physik*, **64**(7–8):563–567, 1930.

[AJL04] J. Ambjørn, J. Jurkiewicz, and R. Loll. Emergence of a 4D World from Causal Quantum Gravity. *Physical Review Letters*, **93**:131301, 2004.

[AJL10a] J. Ambjørn, J. Jurkiewicz, and R. Loll. Deriving Spacetime from First Principles. *Annalen der Physik*, **19**(3–5):186–195, 2010.

[AJL10b] J. Ambjørn, J. Jurkiewicz, and R. Loll. Quantum Gravity, or the Art of Building Spacetime. In D. Oriti, editor, *Approaches to Quantum Gravity*. Cambridge University Press, Cambridge, 2010.

[Alb92] D. Albert. *Quantum Mechanics and Experience*. Harvard University Press, Cambridge, MA, 1992.

[AM73] V. I. Andryushin and V. N. Melnikov. Mossbaüer Effect on Neutrino and Limits on Elementary Length. *Lettere al Nuovo Cimento*, **7**(16):809–810, 1973.

[Ama89] D. Amati. Gravity from Strings. *Physics Reports*, **184**(2–4):105–112, 1989.

[AMPS12] A. Almheiri, D. Marolf, J. Polchinski, and J. Sully. Black Holes: Complementarity or Firewalls? *Preprint*, 2012. arXiv:1207.3123.

[And72] P. Anderson. More is Different. *Science*, **177**(4047):393–396, 1972.

[Ara89a] S. Aramaki. Development of the Renormalization Theory in Quantum Electrodynamics (I). *Historia Scientiarum*, **36**:97–116, 1989.

[Ara89b] S. Aramaki. Development of the Renormalization Theory in Quantum Electrodynamics (II). *Historia Scientiarum*, **37**:91–113, 1989.

[Arn12] F. Arntzenius. *Space, Time, and Stuff*. Oxford University Press, Oxford, 2012.

[ARS92] A. Ashtekar, C. Rovelli, and L. Smolin. Weaving a Classical Metric with Quantum Threads. *Physical Review Letters*, **69**:237–240, 1992.

[AS99] R. Adler and D. Santiago. On Gravity and the Uncertainty Principle. *Modern Physics Letters A*, **14**(20):1371–1381, 1999.

[Ash86] A. Ashtekar. New Variables for Classical and Quantum Gravity. *Physical Review Letters*, **57**:2244–2247, 1986.

[Bai11] J. Bain. Effective Field Theories. In R. Batterman, editor, *The Oxford Handbook of Philosophy of Physics*. Oxford University Press, Oxford, 2011.

[Bai13] J. Bain. The Emergence of Spacetime in Condensed Matter Approaches to Quantum Gravity. *Studies in the History and Philosophy of Modern Physics*, **44**(3):338–345, 2013.

[Ban95] T. Banks. Lectures on Black Holes and Information Loss. *Nuclear Physics B – Proceedings Supplements*, **41**(1–3):21–65, 1995.

[Bar82] J. Barbour. Relational Concepts of Space and Time. *British Journal for the Philosophy of Science*, **33**(3):251–274, 1982.

[Bar89] J. Barbour. *The Discovery of Dynamics*. Oxford University Press, Oxford, 1989.

[Bar00] J. Barbour. *The End of Time: The Next Revolution in Physics*. Oxford University Press, Oxford, 2000.

[Bar09] J. Barbon. Black Holes, Information and Holography. *Journal of Physics – Conference Series*, **171**(1):012009, 2009.

[Bay51] T. Baynes. *The Port-Royal Logic*. Sutherland and Know, Edinburgh, 1851. Translation. Second Edition.

[BC06] M. Braverman and S. Cook. Computing Over the Reals: Foundations for Scientific Computing. *Notices of the AMS*, **53**(3):318–329, 2006.

[BCH73] J. Bardeen, B. Carter, and S. Hawking. The Four Laws of Black Hole Mechanics. *Communications in Mathematical Physics*, **31**:161–170, 1973.

[BCS97] P. Burgisser, M. Claussen, and M. Shokrollahi. *Algebraic Complexity Theory*. Springer, New York, 1997.

[BD65] J. Bjorken and S. Drell. *Relativistic Quantum Fields*. McGraw-Hill, New York, 1965.

[Bek72] J. Bekenstein. Black Holes and the Second Law. *Lettere al Nuovo Cimento*, **4**:737–740, 1972.

[Bek73] J. Bekenstein. Black Holes and Entropy. *Physical Review D*, **7**:2333–2346, 1973.

[Bek74] J. Bekenstein. Generalized Second Law of Thermodynamics in Black-Hole Physics. *Physical Review D*, **9**:3292–3300, 1974.

[Bek03] J. Bekenstein. Information in the Holographic Universe. *Scientific American*, **289**(8):58–65, 2003.

[Bel64] J. Bell. On the Einstein–Podolsky–Rosen Paradox. *Physics*, **1**:195–200, 1964.

[Bel87] J. Bell. How to Teach Special Relativity. In *Speakable and Unspeakable in Quantum Mechanics*, pages 67–80. Cambridge University Press, Cambridge, 1987.

[Bel90] J. Bell. Against Measurement. In A. Miller, editor, *Sixty Two Years of Uncertainty*, pages 17–32. NATO ASI, Plenum Press, New York, 1990.

[Bel92] J. Bell. George Francis Fitzgerald. *Physics World*, **5**:31–35, 1992. Abridged by Denis Weare from a lecture in 1989.

[Bel11] G. Belot. Background Independence. *General Relativity and Gravitation*, **42**(10):2865–2884, 2011.

[Ben62] P. Benacerraf. Tasks, Super-Tasks, and Modern Eleatics. *Journal of Philosophy*, **59**:765–784, 1962.

[Ben03] C. Bennett. Notes on Landauer's Principle, Reversible Computation and Maxwell's Demon. *Studies in the History and Philosophy of Modern Physics*, **34**:501–510, 2003.

[Ber49] P. Bergmann. Non-Linear Field Theories. *Physical Review*, **75**:680–685, 1949.

[BH84] D. Bohm and B. Hiley. Generalization of the Twistor and Clifford Algebras as a Basis for Geometry. In *Revista Brasileira de Fisica, Volume Especial, Os 70 Anos do Mario Schönberg*, pages 1–27. Sociedade Brasileira de Fisica, Sao Paulo, Brasil, 1984.

[Bha39] H. Bhabha. The Fundamental Length Introduced by the Theory of the Mesotron (Meson). Nature, **143**:276–277, 1939.

[BI34] M. Born and L. Infeld. Foundations of the New Field Theory. *Proceedings of the Royal Society of London A*, **144**(852):425–451, 1934.

[BI99] J. Butterfield and C. Isham. Spacetime and the Philosophical Challenge of Quantum Gravity. In C. Callender and N. Huggett, editors, *Physics Meets Philosophy at the Planck Scale*, pages 33–39. Cambridge University Press, Cambridge, 1999.

[BIA05] A. Bassi, E. Ippoliti, and S. Adler. Towards Quantum Superpositions of a Mirror: an Exact Open Systems Analysis. *Physical Review Letters*, **94**:030401, 2005.

[BJ03] Y. Balashov and M. Janssen. Presentism and Relativity. *British Journal for the Philosophy of Science*, **33**:251–274, 2003.

[BKLS86] L. Bombelli, R. Koul, J. Lee, and R. Sorkin. Quantum Source of Entropy for Black Holes. *Physical Review D*, **34**:373–383, 1986.

[Bla50] M. Black. Achilles and the Tortoise. *Analysis*, **11**:91–101, 1950.

[BLMS87] L. Bombelli, J. Lee, D. Meyer, and R. Sorkin. Spacetime As a Causal Set. *Physical Review Letters*, **59**:521–524, 1987.

[BLV11] C. Barcelo, S. Liberati, and M. Visser. Analogue Gravity. *Living Reviews in Relativity*, **14**, 2011. arXiv:gr-gc/0505065v3.

[Boh63] D. Bohm. *Problems in the Basic Concept of Physics*. 1963. An Inaugural Lecture Delivered at Birkbeck College on 13th of February.

[Boh83] N. Bohr. Discussion with Einstein on Epistemological Problems in Atomic Physics. In J. Wheeler and W. Zurek, editors, *Quantum Theory and Measurement*, pages 9–49. Princeton University Press, Princeton, NJ, 1983.

[Bol50] B. Bolzano. *Paradoxes of the Infinite*. Yale University Press, New Haven, CT, 1950.

[Bop40] F. Bopp. Eine lineare Theorie des Elektrons. *Annalen der Physik*, **430**(5):345–384, 1940.

[Bor26] M. Born. Zur Quantenmechanik der Stossvorgänge. *Zeitschrift für Physik*, **37**(12):863–867, 1926.

[Bor33] M. Born. Modified Field Equations with a Finite Radius of the Electron. *Nature*, **132**(3329):282, 1933.

[Bor34] M. Born. Cosmic Rays and the New Field Theory. *Nature*, **133**:63–64, 1934.

[Bor35] M. Born. The mysterious number 137. *Proceedings of the Indian Academy of Science A*, **6**:533–561, 1935.

[Bor38a] M. Born. A Suggestion for Unifying Quantum Theory and Relativity. *Proceedings of the Royal Society of London A*, **165**(921):291–303, 1938.

[Bor38b] M. Born. Application of "Reciprocity" to Nuclei. *Proceedings of the Royal Society of London A*, **166**:552–557, 1938.

[Bor49] M. Born. Reciprocity Theory of Elementary Particles. *Reviews of Modern Physics*, **21**:463–473, 1949.

[Bor50] M. Born. Non-Localizable Fields and Reciprocity. *Nature*, **165**(4190):269–270, 1950.

[Bor53] M. Born. The Conceptual Situation in Physics and the Prospects of its Future Development. *Proceedings of the Royal Society of London A*, **66**:501–513, 1953.

[Bou02] R. Bousso. The Holographic Principle. *Reviews of Modern Physics*, **74**:825–874, 2002.

[Boy49] C. Boyer. *The History of Calculus*. Dover, New York, 1949.

[BP06] H. Brown and O. Pooley. Minkowski Spacetime – a Glorious Non-Entity. In D. Dieks, editor, *The Ontology of Spacetime*, pages 67–92. Elsevier, Amsterdam, 2006.

[BP10] J. Bub and I. Pitowsky. Two Dogmas About Quantum Mechanics. In S. Saunders, J. Barrett, A. Kent, and D. Wallace, editors, *Many Worlds?* Oxford University Press, Oxford, 2010.

[BR33] N. Bohr and L. Rosenfeld. On the Question of the Measurability of Electromagnetic Field Quantities. In R. Cohen and J. Stachel, editors, *The Selected Papers of Leon Rosenfeld*, pages 357–400. Reidel, Dordrecht, 1957/1933.

[BR50] N. Bohr and L. Rosenfeld. Field and Charge Measurements in Quantum Electrodynamics. *Physical Review*, **78**(6):794–798, 1950.

[BR81] H. Brown and M. Redhead. A Critique of the Disturbance Theory of the Indeterminacy of Quantum Mechanics. *Foundations of Physics*, **11**:1–20, 1981.

[BR91] L. Brown and H. Rechenberg. Quantum Field Theories, Nuclear Forces, and Cosmic Rays (1934–1938). *American Journal of Physics*, **59**(7):595–605, 1991.

[Bro36] M. Bronstein. Qúantentheorie Schwacher Gravitationsfelder. *Physikalische Zeitschrift der Sowjetunion*, **9**:140–157, 1936.

[Bro71] J. Bromberg. The Impact of the Neutron: Bohr and Heisenberg. *Historical Studies in the Physical Sciences*, **3**:307–341, 1971.

[Bro93] H. Brown. Correspondence, Invariance and Heuristics in the Emergence of Special Relativity. In S. French and H. Kamminga, editors, *Correspondence, Invariance and Heuristics*, pages 227–260. Kluwer, Amsterdam, 1993.

[Bro05a] H. Brown. Einstein's Misgiving about his 1905 Formulations of STR. *European Journal of Physics*, **26**:S85–S90, 2005.

[Bro05b] H. Brown. *Physical Relativity*. Oxford University Press, Oxford, 2005.

[BS08] Y. Bonder and D. Sudarsky. Quantum Gravity Phenomenology without Lorentz Invariance Violation. *Classical and Quantum Gravity*, **25**:105017, 2008.

[BSH+98] E. Buks, R. Schuster, M. Heiblum, D. Mahalu, and V. Umansky. Dephasing in Electron Interference by a 'Which-path' Detector. *Nature*, **391**:871–874, 1998.

[BSP84] T. Banks, L. Susskind, and M. E. Peskin. Difficulties for the Evolution of Pure States into Mixed States. *Nuclear Physics B*, **244**:125, 1984.

[BT09] M. Blau and S. Theisen. String Theory as a Theory of Quantum Gravity: a Status Report. *General Relativity and Gravitation*, **41**:743–755, 2009.

[Bub97] J. Bub. *Interpreting the Quantum World*. Cambridge University Press, Cambridge, 1997.

[Bub00] J. Bub. Quantum Mechanics as a Principle Theory. *Studies in the History and Philosophy of Modern Physics*, 31(1):75–94, 2000.

[Bur07] C. Burgess. An Introduction to Effective Field Theory. *Annual Review of Nuclear and Particle Science*, 57:329–362, 2007.

[Cal01] C. Callender. Taking Thermodynamics too Seriously. *Studies In the History and Philosophy of Modern Physics*, 32(4):539–553, 2001.

[Cao97] T. Cao. *Conceptual Developments of 20th Century Field Theories*. Cambridge University Press, Cambridge, 1997.

[Car96] C. Carson. The Peculiar Notion of Exchange Forces II: From Nuclear Forces to QED, 1929–1950. *Studies in the History and Philosophy of Modern Physics*, 21(2):99–131, 1996.

[Car00] S. Carlip. Black Hole Entropy from Horizon Conformal Field Theory. *Nuclear Physics Proceedings Supplement*, 88:10–16, 2000.

[Cas81] D. Cassidy. Cosmic Ray Showers, High Energy Physics, and Quantum Field Theories: Programmatic Interactions in the 1930s. *Historical Studies in the Physical Sciences*, 12:1–39, 1981.

[Cas02] E. Castellani. Reductionism, Emergence, and Effective Field Theories. *Studies in the History and Philosophy of Modern Physics*, 33(2):251–267, 2002.

[CH67] P. Cohen and R. Hersh. Non-Cantorian Set Theory. *Scientific American*, 217:104–116, 1967.

[Cha75] G. Chaitin. Randomness and Mathematical Proof. *Scientific American*, 232(5):47–52, 1975.

[Che63] G. F. Chew. The Dubious Role of the Space-Time Continuum in Subatomic Physics. *Science Progress*, 51:529–539, 1963.

[CK95] B. Carazza and H. Kragh. Heisenberg's Lattice World: The 1930 Theory Sketch. *American Journal of Physics*, 63:595, 1995.

[CL83] O. Caldeira and A. Leggett. Quantum Tunneling in a Dissipative system. *Annals of Physics*, 149:374–456, 1983.

[CLS52] E. Courant, M. Livingston, and H. Snyder. The Strong-Focusing Synchroton – A New High Energy Accelerator. *Physical Review*, 88:1190–1196, 1952.

[CM98] C. Callan and J. Maldacena. Brane Dynamics from the Born–Infeld Action. *Nuclear Physics B*, 513:198–212, 1998.

[Coi59] H.R. Coish. Elementary Particles in a Finite World Geometry. *Physical Review*, 114:383–388, 1959.

[Con94] A. Connes. *Noncommutative Geometry*. Academic Press, San Diego, CA, 1994.

[CON10] E. Cobanera, G. Ortiz, and Z. Nussinov. Unified Approach to Quantum and Classical Dualities. *Physical Review Letters*, 104(2):020402, 2010.

[Cop96] J. Copeland. The Church–Turing Thesis. In E. Zalta, editor, *The Stanford Encyclopedia of Philosophy*. 1996.

[Cor05] D. T. Cornwell. Forces Due to Contraction on a Cord Spanning Between Two Spaceships. *Europhysics Letters*, 71(5):699–704, 2005.

[CP53] M. Cheritien and R. Peierls. Properties of Form Factors in Non Local Field Theories. *Il Nuovo Cimento*, 10(5):668–676, 1953.

[CP54] M. Cheritien and R. Peierls. A Study of Gauge-Invariant Non-Local Interactions. *Proceedings of the Royal Society of London A*, 223(1155):468–481, 1954.

[CPS⁺04] J. Collins, A. Perez, D. Sudarsky, L. Urrutia, and H. Vucetich. Lorentz Invariance and Quantum Gravity: an Additional Fine-tuning Problem? *Physical Review Letters*, **93**:191301, 2004.

[CPS10] J. Collins, A. Perez, and D. Sudarsky. Lorentz Invariance Violation and its Role in Quantum Gravity Phenomenology. In D. Oriti, editor, *Approaches to Quantum Gravity*, pages 528–547. Cambridge University Press, Cambridge, 2010.

[CS93] T. Cao and S. Schweber. The Conceptual Foundations and the Philosophical Aspects of Renormalization Theory. *Synthese*, **97**(1):33–108, 1993.

[Cum90] P. Cummins. Bayle, Leibniz, Hume and Reid on Extension, Composites and Simples. *History of Philosophy Quarterly*, **7**(3):299–314, 1990.

[Dar86a] O. Darrigol. The Origin of Quantized Matter Waves. *Historical Studies in the Physical Sciences*, **16**(2):198–253, 1986.

[Dar86b] O. Darrigol. The Quantum Electrodynamical Analogy in Early Nuclear Theory. *Revue d'Histoire des Sciences*, **41**(3–4):225–297, 1986.

[Dar92] O. Darrigol. *From C-numbers to Q-numbers: the Classical Analogy in the History of Quantum Theory*. University of California Press, Berkeley, CA, 1992.

[Dav58] M. Davis. *The Undecidable*. Dover, New York, 1958.

[DB59] E. Dewan and M. Beran. Notes on Stress Effects Due to Relativistic Contraction. *American Journal of Physics*, **27**:517–518, 1959.

[Del04] B. Delamotte. A Hint of Renormalization. *American Journal of Physics*, **72**:170, 2004.

[Des57] S. Deser. General Relativity and the Divergence Problem in Quantum Field Theory. *Reviews of Modern Physics*, **29**(3):417–423, 1957.

[Dew63] E. Dewan. Stress Effects Due to Lorentz Contraction. *American Journal of Physics*, **31**:383–386, 1963.

[DHS04] F. Dowker, J. Henson, and R. Sorkin. Quantum Gravity Phenomenology, Lorentz Invariance and Discreteness. *Modern Physics Letters A*, **19**:1829, 2004.

[Dio89] L. Diosi. Models for Universal Reduction of Macroscopic Quantum Fluctuations. *Physical Review A*, **40**:1165–1174, 1989.

[Dir27] P. Dirac. The Physical Interpretation of Quantum Dynamics. *Proceedings of the Royal Society of London A*, **118**:621–641, 1927.

[Dir29] P. Dirac. Quantum Mechanics of Many Electron Systems. *Proceedings of the Royal Society of London A*, **123**:714–733, 1929.

[Dir38] P. Dirac. Classical Theory of Radiating Electrons. *Proceedings of the Royal Society of London A*, **167**:148–169, 1938.

[Dir60] P. Dirac. A Reformulation of the Born–Infeld Electrodynamics. *Proceedings of the Royal Society A*, **257**:32–43, 1960.

[Dir62] P. Dirac. Particles of Finite Size in the Gravitational Field. *Proceedings of the Royal Society of London A*, **270**:354–356, 1962.

[Dir63] P. Dirac. The Evolution of the Physicist's Picture of Nature. *Scientific American*, **208**(5):45–53, 1963.

[Dir78] P. Dirac. Quantum Electrodynamics. In H. Hora and J. Shepansky, editors, *Directions in Physics*. Wiley, New York, 1978.

[Dir81] P. Dirac. Does Renormalization Make Sense? *Perturbative QCD, AIP Conference Proceedings*, **74**:129–130, 1981.

[DJ13] A. Duncan and M. Janssen. (Never) Mind your p's and q's: Von Nuemann versus Jordan on the Foundations of Quantum Theory. *European Journal of Physics H*, **38**:175–259, 2013.

[DN01] M. Douglas and N. Nekrasov. Noncommutative Field Theory. *Reviews of Modern Physics*, **73**:977–1029, 2001.

[Dou99] D. Dou. *Causal Sets, a Possible Interpretation for the Black Hole Entropy, and Related Topics*. PhD Thesis, SISSA Trieste, 1999.

[DR11] C. DeWitt and D. Rickles, editors. *The Role of Gravitation in Physics, Report from the 1957 Chapel Hill Conference*. Max Planck Research Library for the History and Development of Knowledge, Berlin, 2011.

[DT09] B. Dittrich and T. Thiemann. Are the Spectra of Geometrical Operators in Loop Quantum Gravity Really Discrete? *Journal of Mathematical Physics*, **50**(1):012503–11, 2009.

[Dys49] F. Dyson. The Radiation Theories of Tomonaga, Schwinger, and Feynman. *Physical Review*, **75**:486–502, 1949.

[Ear86] J. Earman. *A Primer on Determinism*. Reidel, Boston, MA, 1986.

[Ear89] J. Earman. *World Enough and Spacetime*. MIT Press, Boston, MA, 1989.

[Ear02] J. Earman. Thoroughly Modern Mctaggart: or, What Mctaggart Would Have Said If He Had Read the General Theory of Relativity. *Philosophers' Imprint*, **2**(3):1–28, 2002.

[Ear03a] J. Earman. The Cosmological Constant, the Fate of the Universe, Unimodular Gravity, and All That. *Studies in the History and Philosophy of Modern Physics*, **34**(4):559–577, 2003.

[Ear03b] J. Earman. Tracking Down Gauge: an Ode to the Constrained Hamiltonian Formalism. In K. Brading and E. Castellani, editors, *Symmetries in Physics: Philosophical Reflections*. Cambridge University Press, Cambridge, 2003.

[Ear04] J. Earman. Determinism: What We Have Learned and What We Still Don't Know. In J. Campbell, M. O'Rourke, and D. Shier, editors, *Freedom and Determinism, Topics in Contemporary Philosophy Series, Vol. 2*. Seven Springs Press, 2004.

[EB99] J. Earman and G. Belot. Presocratic Quantum Gravity. In C. Callender and N. Huggett, editors, *Physics Meets Philosophy at the Planck Scale*, pages 213–255. Cambridge University Press, Cambridge, 1999.

[Edd23] A. Eddington. *The Mathematical Theory of Relativity*. Cambridge University Press, Cambridge, 1923.

[Edd28] A. Eddington. *The Nature of the Physical World*. Cambridge University Press, Cambridge, 1928.

[Edd41] A. Eddington. Philosophy of Physical Science. *Nature*, **148**:692–693, 1941.

[Edd46] A. Eddington. *Fundamental Theory*. Cambridge University Press, Cambridge, 1946.

[Efi72] G. Efimov. On the Construction of Nonlocal Quantum Electrodynamics. *Annals of Physics*, **71**:466–485, 1972.

[EH77] K. Eppley and E. Hannah. The Necessity of Quantizing the Gravitational Field. *Foundations of Physics*, **7**:51–68, 1977.

[EHNS84] J. Ellis, J. Hagelin, D. Nanopoulos, and M. Srednicki. Search for Violations of Quantum Mechanics. *Nuclear Physics B*, **241**(2):381–405, 1984.

[Ein15] A. Einstein. Theoretische Atomistik. In *Die Kultur der Gegenwart. Ihre Entwicklung und ihre Ziele. Part 3, Mathematik, Naturwissenschaften, Medizin. Section 3, Anorganischen Naturwissenschaften*. 1915.

[Ein16] A. Einstein. Naherungsweise Integration der Feldgleichungen der Gravitation. *Koniglich Preussische Akademie der Wissenschaften, (Berlin) Sitzungsberichte*, pages 688–696, 1916.

[Ein21] A. Einstein. Geometrie und Erfahrung. 1921. Reprinted as Doc. 52 in M. Janssen, *et al.*, editors, *The Collected Papers of Albert Einstein, Vol. 7. The Berlin Years: Writings, 1918–1921*. Princeton University Press, Princeton, NJ, 2002. Translation (Geometry and experience) in A. Einstein, *Ideas and Opinions*. Crown, New York, 1954.

[Ein29] A. Einstein. The History of Field Theory (The Olds and News of Field Theory). *The New York Times*. February 3, 1929.

[Ein54a] A. Einstein. What is the Theory of Relativity? In *Ideas and Opinions*, pages 232–237. Crown, New York, 1919/1954.

[Ein54b] A. Einstein. *Ideas and Opinions*. Crown, New York, 1954.

[Ein89a] A. Einstein. Bemerkung zur Notiz des Haerrn P. Ehrenfest, Document 47. In M. Klein, editor, *The Collected Papers of Albert Einstein, Vol. 5. The Swiss Years: Writings, 1900–1909 (English Supplement)*, pages 236–237. Princeton University Press, Princeton, NJ, 1907/1989.

[Ein89b] A. Einstein. Letter to Sommerfeld, Document 73. In M. Klein, editor, *The Collected Papers of Albert Einstein, Vol. 5. The Swiss Years: Correspondence, 1902–1914 (English Supplement)*. Princeton University Press, Princeton, NJ, 1908/1989.

[EMN99] J. Ellis, N. Mavromatos, and D. Nanopoulos. Search for Quantum Gravity. *General Relativity and Gravitation*, **31**:1257–1262, 1999.

[Eyg65] L. Eyges. Physics of the Mössbauer Effect. *American Journal of Physics*, **33**:790–802, 1965.

[Fal07] B. Falkenburg. *Particle Metaphysics*. Springer, Berlin, 2007.

[Fey48a] R. Feynman. A Relativistic Cut-Off for Classical Electrodynamics. *Physical Review*, **74**:939–946, 1948.

[Fey48b] R. Feynman. A Relativistic Cut-Off for Quantum Electrodynamics. *Physical Review*, **74**:1430–1438, 1948.

[Fey65] R. Feynman. *The Character of Physical Law*. MIT Press, Cambridge, MA, 1965.

[FH65] R. Feynman and A. R. Hibbs. *Quantum Mechanics and Path Integrals*. Dover Publications, New York, 1965.

[Fin49] R. Finkelstein. On the Quantization of a Unitary Field Theory. *Physical Review*, **75**:1079–1087, 1949.

[Fin86] A. Fine. *The Shaky Game*. University of Chicago Press, Chicago, IL, 1986.

[Fit79] J. Fitzgerald. *Alfred North Whitehead's Early Philosophy of Space and Time*. University Press of America, Washington, DC, 1979.

[FKGN07] L. Freidel, J. Kowalski-Glikman, and S. Nowak. From Noncommutative κ-Minkowski to Minkowski Spacetime. *Physics Letters B*, **648**(1):70–75, 2007.

[Fli28] H. Flint. Relativity and the Quantum Theory. *Proceedings of the Royal Society of London A*, **117**(778):630–637, 1928.

[Fli36] H. Flint. On the Development of the Quantum Equation and a Possible Limit to its Application. *Proceedings of the Physical Society*, **48**(3):433–444, 1936.

[Fli38] H. Flint. The Ratio of the Masses of the Fundamental Particles. *Proceedings of the Physical Society*, **50**(1):90, 1938.

[FMW03] R. Feynman, F. Moringo, and W. Wagner. *Feynman Lectures on Gravitation*. Westview Press, Cambridge, MA, 2003.

[Fog88] R. Fogelin. Hume and Berkeley on the Proofs of Infinite Divisibility. *Philosophical Review*, **97**(1):47–69, 1988.

[For83] J. Ford. How Random is a Coin Toss. *Physics Today*, **4**:40–47, 1983.

[For95] P. Forrest. Is Spacetime Discrete or Continuous? – An Empirical Question. *Synthese*, **103**(3):327–354, 1995.

[Fox08] T. Fox. Haunted by the Spectre of Virtual Particles: a Philosophical Reconsideration. *Journal for the General Philosophy of Science*, **39**:35–51, 2008.

[Fra08] D. Fraser. The Fate of Particles in Quantum Field Theories with Interactions. *Studies in the History and Philosophy of Modern Physics*, **39**(4):841–859, 2008.

[Fra09] D. Fraser. Quantum Field Theory: Underdetermination, Inconsistency, and Idealization. *Philosophy of Science*, **76**(4):536–567, 2009.

[Fra11] D. Fraser. How to Take Particle Physics Seriously: a further Defense of Axiomatic Quantum Field Theory. *Studies in the History and Philosophy of Modern Physics*, **42**(2):126–135, 2011.

[Fre90] E. Fredkin. Digital Mechanics: an Informational Process Based on Reversible Universal Cellular Automata. *Physica D*, **45**:254–270, 1990.

[Fri83] M. Friedman. *Foundations of Spacetime Theories*. Princeton University Press, Princeton, NJ, 1983.

[Fri02] M. Friedman. Geometry as a Branch of Physics. In D. Malament, editor, *Reading Natural Philosophy*, pages 193–229. Open Court, Chicago, IL, 2002.

[FW53] H. Flint and E. Williamson. Coordinate Operators and Fundamental Lengths. *Physical Review*, **90**:318–319, 1953.

[FW98] K. Ford and J. A. Wheeler. *Geons, Black Holes, and Quantum Foam*. W.W. Norton, New York, 1998.

[Gal83] P. Galison. The Discovery of the Muon and the Failed Revolution against Quantum Electrodynamics. *Centaurus*, **26**:262–316, 1983.

[Gal87] P. Galison. *How Experiments End*. University of Chicago Press, Chicago, IL, 1987.

[Gal99] G. Gallavotti. *Statistical Mechanics*. Springer-Verlag, Berlin, 1999.

[Gam62] G. Gamow. *Gravity – Classical and Modern Views*. Dover, New York, 1962.

[Gan80] R. Gandy. Church's Thesis and the Principles for Mechanisms. In H. Barwise, H. Keisler, and K. Kunen, editors, *The Kleene Symposium*, pages 123–148. North Holland, 1980.

[Gar95] L. Garay. Quantum Gravity and Minimum Length. *International Journal of Modern Physics A*, **10**:145–165, 1995.

[GDZ92] S. Goldstein, D. Durr, and N. Zanghi. Quantum Equilibrium and the Origin of Absolute Uncertainty. *Journal of Statistical Physics*, **67**(5–6):843–907, 1992.

[Geo93] H. Georgi. Effective Field Theory. *Annual Review of Nuclear and Particle Science*, **43**:209–252, 1993.

[GF94] G. Gorelik and V. Frenkel. *Matvei Petrovich Bronstein and Soviet Theoretical Physics in the Thirties*. Birkhäuser, Basel, 1994.

[Ghi11] G. Ghirardi. Collapse Theories. In E. Zalta, editor, *The Stanford Encyclopedia of Philosophy*. Winter 2011 edition, 2011.

[GL05] F. Girelli and E. Livine. Physics of Deformed Special Relativity: Relativity Principle Revisited. *Brazilian Journal of Physics*, **35**:432–438, 2005.

[Gor92] G. Gorelik. First Steps of Quantum Gravity and the Planck Values. In J. Eisenstaedt and A. Kox, editors, *Studies in the History of General Relativity [Einstein Studies. Vol. 3]*, pages 364–379. Birkhäuser, Basel, 1992.

[Gor05] G. Gorelik. Matvei Bronstein and Quantum Gravity: 70th Anniversary of the Unsolved Problem, *Physics Uspekhi*, **48**:1039, 2005.

[GP99] R. Gambini and J. Pullin. Nonstandard Optics from Quantum Spacetime. *Physical Review D*, **59**:124021, 1999.

[Grü52] A. Grünbaum. A Consistent Conception of the Extended Linear Continuum as an Aggregate of Unextended Elements. *Philosophy of Science*, **19**(4):288–306, 1952.

[Grü53] A. Grünbaum. Whitehead's Method of Extensive Abstraction. *British Journal of the Philosophy of Science*, **4**(15):215–226, 1953.

[GRW80] G. Ghirardi, A. Rimini, and T. Weber. A General Argument Against Superluminal Transmission through the Quantum Mechanical Measurement Process. *Lettere al Nuovo Cimento*, **27**(10):293–298, 1980.

[GRW86] G. Ghirardi, A. Rimini, and T. Weber. Unified Dynamics for Microscopic and Macroscopic Systems. *Physical Review D*, **34**(2):470–491, 1986.

[GV85] G. Gerla and R. Volpe. Geometry Without Points. *American Mathematical Monthly*, **92**(10):707–711, 1985.

[Hag06] A. Hagar. Quantum Computing. In E. Zalta, editor, *The Stanford Encyclopedia of Philosophy*. 2006.

[Hag07] A. Hagar. Experimental Metaphysics$_2$. *Studies in the History and Philosophy of Modern Physics*, **38**(4):906–919, 2007.

[Hag08] A. Hagar. Length Matters: the Einstein–Swann Correspondence. *Studies in the History and Philosophy of Modern Physics*, **39**(3):532–556, 2008.

[Hag09] A. Hagar. Minimal Length in Quantum Gravity and the Fate of Lorentz Invariance. *Studies in the History and Philosophy of Modern Physics*, **40**(3):259–267, 2009.

[Hag10] A. Hagar. *The Complexity of Noise: A Philosophical Outlook on Quantum Error Correction*. Morgan & Claypul Publishing, 2010.

[Hag12] A. Hagar. Decoherence: the View from the History and the Philosophy of Science. *Philosophical Transactions of the Royal Society of London A*, **370**(1975):4594–4609, 2012.

[Haw71] S. Hawking. Gravitational Radiation from Colliding Black Holes. *Physical Review Letters*, **26**:1344–1346, 1971.

[Haw75] S. Hawking. Particle Creation by Black Holes. *Communications in Mathematical Physics*, **43**(3):199–220, 1975.

[Haz90] A. Hazen. The Mathematical Philosophy of Contact. *Philosophy*, **65**:205–211, 1990.

[HC99] N. Huggett and C. Callender. *Physics Meets Philosophy at the Planck Scale*. Cambridge University Press, Cambridge, 1999.

[Hea02] R. Healey. Can Physics Coherently Deny the Reality of Time? *Royal Institute of Philosophy Supplements*, **50**:293–316, 2002.

[Hei27] W. Heisenberg. The Physical Content of Quantum Kinematics and Mechanics. In J. Wheeler and W. Zurek, editors, *Quantum Theory and Measurement*, pages 62–84. Princeton University Press, Princeton, NJ, 1983/1927.

[Hei30a] W. Heisenberg. *The Physical Principles of Quantum Mechanics*. Dover Publications, New York, 1930.

[Hei30b] W. Heisenberg. The Self-Energy of the Electron. In A. Miller, editor, *Early Quantum Electrodynamics*, pages 121–128. Cambridge University Press, Cambridge, 1994/1930.

[Hei36] W. Heisenberg. Zur Theorie der 'Schauer' in der Höhenstrahlung. *Zeitschrift für Physik*, **101**:533–540, 1936.

[Hei38a] W. Heisenberg. Die Grenzen der Anwendbarkeit der bisherigen Quanten-theorie. *Zeitschrift für Physik*, **110**:251–266, 1938.

[Hei38b] W. Heisenberg. The Universal Length Appearing in the Theory of Elementary Particles. *Annalen der Physik*, **32**:20–33, 1938.

[Hei43] W. Heisenberg. Die beobachtbaren Grössen in der Theorie der Elementarteilchen. *Zeitschrift für Physik*, **120**:513–538; 673–702, 1943.

[Hei51a] W. Heisenberg. On the Mathematical Frame of the Theory of Elementary Particles. *Communications on Pure and Applied Mathematics*, **4**(1):15–22, 1951.

[Hei51b] W. Heisenberg. Paradoxien des Zeitbegriffs in der Theorie der Elementarteilchen. In M. Born *et al.* editors, *Festschrift zur Feier des Zweihundertjährigen Bestehens der Akademie der Wissenschaften in Göttingen*, pages 50–64. Springer-Verlag, Berlin, 1951.

[Hei62] W. Heisenberg. *Physics and Philosophy*. Harper, New York, 1962.

[Hem45] C. Hempel. Geometry and Empirical Science. *American Mathematical Monthly*, **52**:7–17, 1945.

[Hen10] J. Henson. The Causal Set Approach to Quantum Gravity. In D. Oriti, editor, *Approaches to Quantum Gravity*, pages 393–413. Cambridge University Press, Cambridge, 2010.

[HH13] A. Hagar and M. Hemmo. The Primacy of Geometry. *Studies in the History and Philosophy of Modern Physics*, **44**(3):357–364, 2013.

[Hil25] D. Hilbert. On the Infinite. In P. Benacerraf and H. Putnam, editors, *The Philosophy of Mathematics*, pages 183–201. Cambridge University Press, Cambridge, 1983/1925.

[Hil55] E. Hill. Relativistic Theory of Discrete Momentum Space and Discrete Spacetime. *Physical Review*, **100**:1780–1783, 1955.

[HK91] R. Hovis and H. Kragh. Resource Letter HEPP-1: History of Elementary-Particle Physics. *American Journal of Physics*, **59**(9):779–807, 1991.

[Hog94] M. Hogarth. Non-Turing Computers and Non-Turing Computability. *PSA*, **1**:126–138, 1994.

[Hol72] G. Holton. On Trying to Understand Scientific Genius. *American Scholar*, **41**:95–110, 1972.

[Hor98] G. Horowitz. Quantum States of Black Holes. In R. Wald, editor, *Black Holes and Relativistic Stars*, pages 241–266. University of Chicago Press, Chicago, IL, 1998.

[Hor05] G. Horowitz. Spacetime in String Theory. *New Journal of Physics*, **7**:201, 2005.

[Hos06] S. Hossenfelder. Interpretation of Quantum Field Theories with a Minimal Length Scale. *Physical Review D*, **73**:105013, 2006.

[Hos07] S. Hossenfelder. Multiparticle States in Deformed Special Relativity. *Physical Review D*, **75**:105005, 2007.

[Hos09] S. Hossenfelder. The Box-Problem in Deformed Special Relativity. *Preprint*, 2009. arXiv:0912.0090.

[Hos10] S. Hossenfelder. Bounds on an Energy-Dependent and Observer-Independent Speed of Light from Violations of Locality. *Physical Review Letters*, **104**:140402, 2010.

[Hos12] S. Hossenfelder. Quantum Superpositions of the Speed of Light. *Foundations of Physics*, **42**:1452, 2012.

[Hos13] S. Hossenfelder. Minimal Length Scale Scenarios for Quantum Gravity. *Living Reviews in Relativity*, **16**(2), 2013.

[HOST13a] A. Hanson, G. Ortiz, A. Sabry, and Y. Tai. Discrete Quantum Theories. *Preprint*, 2013. arXiv:1305.3292.

[HOST13b] A. Hanson, G. Ortiz, A. Sabry, and Y. Tai. Geometry of Discrete Quantum Computing. *Journal of Physics A: Mathematical and Theoretical*, **46**(18):185301, 2013.

[How85] D. Howard. Einstein on Locality and Separability. *Studies in the History and the Philosophy of Science*, **16**:171–201, 1985.

[How04] D. Howard. Einstein's Philosophy of Science. *Stanford Encyclopedia of Philosophy*, 2004. Available at http://plato.stanford.edu/entries/einstein-philscience/.

[HP29] W. Heisenberg and W. Pauli. Zur Quantendynamik der Wellenfelder. *Zeitschrift für Physik*, **56**:1–61, 1929.

[HS33] W. Heitler and F. Sauter. Stopping of Fast Particles with Emission of Radiation and the Birth of Positive Electrons. *Nature*, **132**:892–893, 1933.

[HS06] M. Hemmo and O. Shenker. Von Neumann's Entropy does not Correspond to Thermodynamic Entropy. *Philosophy of Science*, **73**:153–174, 2006.

[HS10] M. Hemmo and O. Shenker. Maxwell's Demon. *Journal of Philosophy*, **107**(8):389–411, 2010.

[HS11] A. Hagar and G. Sergioli. Counting Steps: a New Interpretation of Objective Chance in Statistical Physics. *Preprint*, 2011. arXiv:1101.3521.

[HT90] L. Hughston and K. Tod. *An Introduction to General Relativity*. Cambridge University Press, Cambridge, 1990.

[HU83] J. Hilgevoord and J. Uffink. Overall Width, Mean Peak Width, and the Uncertainty Principle. *Physics Letters A*, **95**(9):474–476, 1983.

[HU88a] J. Hilgevoord and J. Uffink. Interference and Distinguishability in Quantum Mechanics. *Physica B*, **151**:309, 1988.

[HU88b] J. Hilgevoord and J. Uffink. The Mathematical Expression of the Uncertainty Principle. In A. van der Merwe, F. Selleri, and G. Tarozzi, editors, *Microphysical Reality and Quantum Description*, pages 91–114. Kluwer, Dordrecht, 1988.

[HU89] J. Hilgevoord and J. Uffink. Spacetime Symmetries and the Uncertainty Principle. *Nuclear Physics B (Proc. Sup.)*, **6**:246–248, 1989.

[HU90] J. Hilgevoord and J. Uffink. A New View on the Uncertainty Principle. In A. E. Miller, editor, *Sixty-Two years of Uncertainty, Historical and Physical Inquiries into the Foundations of Quantum Mechanics*, pages 121–139. Plenum, New York, 1990.

[Hug09] N. Huggett. Understanding Spacetime. *Philosophy of Science*, **76**(3):404–422, 2009.

[HW41] W. Hurewicz and H. Wallman. *Dimension Theory*. Princeton University Press, Princeton, NJ, 1941.

[HW95] N. Huggett and R. Weingard. The Renormalisation Group and Effective Field Theories. *Synthese*, **102**(1):171–194, 1995.

[Inf64] L. Infeld. *Relativistic Theories of Gravitation*. Pergamon Press, Oxford, 1964.

[Ish95] C. Isham. Structural Issues in Quantum Gravity. Technical Report IMPERIAL/TP/95–96/07, Imperial College, 1995. Writeup of GR14 Plenary Lecture. See also http://arxiv.org/abs/gr-qc/9510063.

[Jac95] T. Jacobson. Thermodynamics of Spacetime: the Einstein Equation of State. *Physical Review Letters*, **75**:1260–1263, 1995.

[Jac10] A. Jacobsen. Crisis, Measurement Problems, and Controversy in Early Quantum Electrodynamics. In C. Carson, A. Kojevnikov, and H. Trischler, editors, *Weimar Culture and Quantum Mechanics*, pages 375–396. Imperial College Press, London, 2010.

[Jam74] M. Jammer. *The Philosophy of Quantum Mechanics*. John Wiley & Sons, New York, 1974.

[Jan95] M. Janssen. *A Comparison between Lorenz's Ether Theory and Special Relativity in the Light of the Experiments of Trouton and Noble*. PhD Thesis, University of Pittsburgh, 1995.

[Jan00] M. Janssen. The Old Sage vs. the Young Turk, 2000. Talk slides, available at http://netfiles.umn.edu/users/janss011/homepage/sage.pdf.

[Jan09] M. Janssen. Drawing the Line between Kinematics and Dynamics in Special Relativity. *Studies in the History and Philosophy of Modern Physics*, **40**(1):26–52, 2009.

[Jar89] J. Jarret. Bell's Theorems: A Guide to the Implications. In J. Cushing and E. Mcmullin, editors, *Philosophical Consequences of Quantum Theory*, pages 60–79. University of Notre Dame Press, Notre Dame, IN, 1989.

[Jea41] J. H. Jeans. The Philosophy of Physical Science. *Nature*, **148**:140, 1941.

[JM07] M. Janssen and M. Mecklenburg. From Classical to Relativistic Mechanics: Electromagnetic Models of the Electron. In V. Hendricks *et al.*, editors, *Interactions: Mathematics, Physics and Philosophy, 1860–1930*, pages 65–134. Springer, Dordrecht, 2007.

[JV03] S. Judes and M. Visser. Conservation Laws in Doubly Special Relativity. *Physical Review D*, **68**:045001, 2003.

[Kad66] L. Kadanoff. Scaling Laws for Ising Models near T_c. *Physics*, **2**:263, 1966.

[Kal71] J. Kalckar. Measurability Problems in the Quantum Theory of Fields. In B. d'Espagnant, editor, *Rendiconti della Scuola Internationale di Fisica: Fundamenti di Meccanica Quantistica*, pages 127–169. Academic Press, New York, 1971.

[Kan02] I. Kant. *Prolegomena to Any Future Metaphysics*. Open Court, Chicago, IL, 1783/1902. Translated by P. Carus.

[KC94] H. Kragh and B. Carazza. From Time Atoms to Spacetime Quantization: the Idea of Discrete Time, ca 1925–1936. *Studies in the History and Philosophy of Science*, **25**(3):437–462, 1994.

[KGN03] J. Kowalski-Glikman and S. Nowak. Doubly Special Relativity and de Sitter Space. *Classical and Quantum Gravity*, **20**:4799–4816, 2003.

[Kie07] C. Kiefer. *Quantum Gravity*. Oxford University Press, Oxford, 2007.

[Kir62] D. Kirzhnits. Field Theory with Nonlocal Interaction. I. Construction of the Unitary S-Matrix. *Soviet Physics JETP*, **14**(2):395–400, 1962.

[Kir64] D. Kirzhnits. Field Theory with Nonlocal Interaction II. The Dynamical Apparatus of the Theory. *Soviet Physics JETP*, **18**(1):103–110, 1964.

[Kir67] D. Kirzhnits. Nonlocal Quantum Field Theory. *Soviet Physics Uspekhi*, **9**:692–700, 1967.

[Kle56] O. Klein. Generalizations of Einstein's Theory of Gravitation Considered from the Point of View of Quantum Field Theory. In A. Mercier and M. Kervaire, editors, *Jubilee of Relativity Theory*, pages 58–71. Birkhauser, Basel, 1956.

[Kle57] O. Klein. Some Remarks on GTR and the Divergence Problem of QFT. *Il Nuovo Cimento Supplement*, **6**(1):344–348, 1957.

[Kle67] M. Klein. Thermodynamics in Einstein's Thought. *Science*, **157**:509–516, 1967.

[KM87] A. Kline and C. Matheson. The Logical Impossibility of Collision. *Philosophy*, **62**(242):509–515, 1987.

[KMM04] D. Kimberly, J. Magueijo, and J. Medeiros. Non-Linear Relativity in Position Space. *Physical Review D*, **70**:084007, 2004.

[Kos04] A. Kostelecký. Gravity, Lorentz Violation, and the Standard Model. *Physical Review D*, **69**:105009, 2004.

[Kra92a] H. Kragh. Relativistic Collisions: The Work of Christian Möller in the Early 1930s. *Archive for the History of Exact Sciences*, **43**:299–328, 1992.

[Kra92b] H. Kragh. Unifying Quanta and Relativity? In M. Bitbol and O. Darigol, editors, *Erwin Schrödinger: Philosophy and the Birth of Quantum Mechanics*, pages 315–337. Frontiérs, Paris, 1992.

[Kra95] H. Kragh. Arthur March, Werner Heisenberg, and the Search for a Smallest Length. *Revue d'Histoire des Sciences (Paris)*, **48**(44):401–434, 1995.

[KS67] S. Kochen and E. Specker. The Problem of Hidden Variables in Quantum Mechanics. *Journal of Mathematics and Mechanics*, **17**:59–87, 1967.

[Kuh78] Thomas Kuhn. *Blackbody Theory and the Quantum Discontinuity, 1894–1912*. University of Chicago Press, Chicago, IL, 1978.

[L$^+$87] A. Leggett *et al.* Dynamics of the Dissipative Two-State System. *Reviews of Modern Physics*, **59**(1):1–85, 1987.

[Lan02] M. Lange. *An Introduction to the Philosophy of Physics*. Blackwell Publishers, Oxford, 2002.

[Lar09] J. Laraudogoitia. Supertasks. In E. Zalta, editor, *The Stanford Encyclopedia of Philosophy*. 2009.

[Lau90] L. Laudan. Demystifying Underdetermination. In C. Savage, editor, *Scientific Theories*, volume 14 of *Minnesota Studies in the Philosophy of Science*, pages 267–297. University of Minnesota Press, Minneapolis, MN, 1990.

[Lau03] R. Laughlin. Emergent Relativity. *International Journal of Modern Physics A*, **18**(6):831–853, 2003.

[Lee36] H. Lee. *Zeno of Elea*. Cambridge University Press, Cambridge, 1936.

[Lez66] A. Leznov. Macrocausality Condition in Nonlocal Field Theories. *Soviet Physics JETP*, **22**:545–546, 1966.

[Lin63] J. Linsley. Evidence for a Primary Cosmic-Ray Particle with Energy 10^{20} eV. *Physical Review Letters*, **10**:146–148, 1963.

[LK65] A. Leznov and D. Kirzhnits. Field Theory with Nonlocal Interaction IV. Questions of Convergence, Causality, and Gauge Invariance. *Soviet Physics JETP*, **21**(2):411–417, 1965.

[Lor16] H. Lorentz. *Theory of the Electron*. Dover, New York, 1916.

[LP31] L. Landau and R. Peierls. Extensions of the Uncertainty Principle to Relativistic Quantum Theory. In J. Wheeler and W. Zurek, editors, *Quantum Theory and Measurement*, pages 465–476. Princeton University Press, Princeton, NJ, 1983/1931.

[LS74] C. Lin and L. Segal. *Mathematics Applied to Deterministic Problems in the Natural Sciences*. MacMillan, New York, 1974.

[Lyn71] A. Lynn. New Models for the Real Number Line. *Scientific American*, **224**:92–99, 1971.

[Mag93] M. Maggiore. A Generalized Uncertainty Principle in Quantum Gravity. *Physics Letters B*, **304**:65–69, 1993.

[Mag03] J. Magueijo. New Varying Speed of Light Theories. *Reports on Progress in Physics*, **66**(11):2025–2068, 2003.

[Mal98] J. Maldacena. The Large *N* Limit of Superconformal Field Theories and Supergravity. *Advances in Theoretical and Mathematical Physics*, **2**:231–252, 1998.

[Mar36] A. March. Die Geometrie Kleinster Röume. *Zeitschrift für Physik*, **104**:161–168, 1936.

[Mar51] A. March. *Quantum Mechanics of Particles and Wave Fields*. John Wiley, New York, 1951.

[Mar73] R. Marnelius. Action Principle and Nonlocal Field Theories. *Physical Review D*, **8**:2472, 1973.

[Mar74] R. Marnelius. Can the *S*-Matrix be Defined in Relativistic Quantum Field Theories with Nonlocal Interaction? *Physical Review D*, **10**:3411, 1974.

[Mat05] D. Mattingly. Modern Tests of Lorentz Invariance. *Living Reviews in Relativity*, **8**:5, 2005.

[Mat06] J. Mattingly. Why Eppley and Hannah's Thought Experiment Fails. *Physical Review D*, **73**:064025, 2006.

[Mat08] D. Mattingly. Have We Tested Lorentz Invariance Enough? *arXiv*: gr-qc/0802.1561v1, 2008.

[Mau94] T. Maudlin. *Quantum Non Locality and Relativity*. Blackwell, Oxford, 1994.

[Mau96] T. Maudlin. Spacetime in the Quantum World. In J. Cushing *et al.*, editors, *Bohmian Mechanics and Quantum Theory: an Appraisal*, pages 285–307. Kluwer Academic Publishers, Boston, MA, 1996.

[Mau01] T. Maudlin. Interpreting Probabilities: What's Interference Got to Do with It? In J. Bricmont *et al.*, editors, *Chance in Physics*, volume 571, pages 283–288. Springer Lecture Notes in Physics, Berlin, 2001.

[Mau07] T. Maudlin. What Could be Objective about Probabilities? *Studies in the History and Philosophy of Modern Physics*, **38**:275–291, 2007.

[McK60] W. McKinley. Search for a Fundamental Length in Microscopic Physics. *American Journal of Physics*, **28**(2):129–134, 1960.

[McM48] H. McManus. Classical Electrodynamics without Singularities. *Proceedings of the Royal Society of London A*, **195**:323–336, 1948.

[Mea64] C. Mead. Possible Connection Between Gravitation and Fundamental Length. *Physical Review*, **135**:849, 1964.

[Mea66] C. Mead. Observable Consequences of Fundamental-Length Hypotheses. *Physical Review*, **143**:990–1005, 1966.

[Mea01] C. Mead. Walking the Planck Length through History. *Physics Today*, **54**:15, 2001.

[Mer05] N. D. Mermin. *It's About Time*. Princeton University Press, Princeton, NJ, 2005.

[Mes61] A. Messiah. *Quantum Mechanics*, Volume II. Interscience Publishers, New York, 1961.

[Mey99] D. Meyer. Finite Precision Measurement Nullifies the Kochen–Specker Theorem. *Physical Review Letters*, **83**:3751–3754, 1999.

[Mie12] G. Mie. Grundlagen einer Theorie der Materie. *Annalen der Physik*, **37**:511, 1912.

[Mil94] A. Miller. *Early Quantum Electrodynamics*. Cambridge University Press, Cambridge, 1994.

[Mis57] C. Misner. Feynman Quantization of General Relativity. *Reviews of Modern Physics*, **29**:497–509, 1957.

[Moo90] C. Moore. Unpredictability and Undecidability in Dynamical Systems. *Physical Review Letters*, **64**(20):2354–2357, 1990.

[Moo96] C. Moore. Recursion Theory on the Reals and Continuous-Time Computation. *Theoretical Computer Science*, **162**(1):23–44, 1996.

[Mor74] H. Morris. The Present Status of the Coish Model. *International Journal of Theoretical Physics*, **9**(6):369–377, 1974.

[Mos62] R. Mössbauer. Nuclear Resonance Fluorescence of Gamma Radiation in Ir^{191}. In H. Frauenfelder, editor, *The Mossbaüer Effect*, pages 101–126. W.A. Benjamin, New York, 1958/1962.

[Moy49] E. Moyal. Quantum Mechanics as a Statistical Theory. *Proceedings of the Cambridge Philosophical Society*, **45**:99–124, 1949.

[Moy12] M. Moyer. Is Space Digital? *Scientific American*, **306**(2):30–37, 2012.

[MS02] J. Magueijo and L. Smolin. Lorentz Invariance with an Invariant Energy Scale. *Physical Review Letters*, **88**:190403, 2002.

[MS03] J. Magueijo and L. Smolin. Generalized Lorentz Invariance with an Invariant Energy Scale. *Physical Review D*, **67**:44017, 2003.

[MS07] F. Markopoulou and L. Smolin. Disordered Locality in Loop Quantum Gravity States. *Classical and Quantum Gravity*, **24**:3813–3824, 2007.

[MSPB03] W. Marshall, C. Simon, R. Penrose, and D. Bouwmeester. Towards Quantum Superpositions of a Mirror. *Physical Review Letters*, **91**:130401, 2003.

[Muk11] S. Mukhi. String Theory: a Perspective Over the Last 25 Years. *Classical and Quantum Gravity*, **28**(15):153001, 2011.

[N+09] H. Nishino *et al.* Search for Proton Decay in a Large Water Cherenkov Detector. *Physical Review Letters*, **102**:141801, 2009.

[NC00] M. Nielsen and I. Chuang. *Quantum Computation and Quantum Information*. Cambridge University Press, Cambridge, 2000.

[OS39] J. Oppenheimer and H. Snyder. On Continued Gravitational Contraction. *Physical Review*, **56**:455–459, 1939.

[Ove13] D. Overbye. A Black Hole Mystery Wrapped in a Firewall Paradox. *The New York Times*, August 12, 2013.

[Owe01] G. Owen. Zeno and the Mathematicians. In W. Salmon, editor, *Zeno's Paradoxes*, pages 139–163. Hacket, Indianapolis, IN, 1957/2001.

[Pau21] W. Pauli. *Theory of Relativity*. Dover, New York, 1921.

[Pau53] W. Pauli. On the Hamiltonian Structure of Nonlocal Field Theories. *Il Nuovo Cimento*, **10**(5):648–667, 1953.

[Pau79] W. Pauli. In K. Meyenn, editor, *Wissenschaftlicher Briefwechsel mit Bohr, Einstein, Heisenberg. Band I: 1919–1929*. Springer Verlag, Berlin, 1979.

[Pau05] W. Pauli. In K. Meyenn, editor, *Wissenschaftlicher Briefwechsel mit Bohr, Einstein, Heisenberg. Band III*. Springer Verlag, Berlin, 2005.

[Pea86] P. Pearle. Suppose the State Vector is Real. In D. Greenberger, editor, *New Techniques and Ideas in Quantum Measurement Theory*, pages 539–554. New York Academy of Science, New York, 1986.

[Pei63] R. Peierls. Field Theory Since Maxwell. In C. Domb, editor, *Clerk Maxwell and Modern Science*, pages 26–42. Athelone Press, University of London, 1963.

[Pei91] R. Peierls. In Defense of 'Measurement'. *Physics Today*, **1**:19–20, 1991.

[Pen00] R. Penrose. Wavefunction Collapse as a Real Gravitational Effect. In A. Fokas *et al.*, editors, *Mathematical Physics 2000*, pages 266–282. Imperial College Press, London, 2000.

[PeR81] M. Pour-el and I. Richards. The Wave Equation with Computable Initial Data such that its Unique Solution is not Computable. *Advances in Mathematics*, **39**:215–239, 1981.

[PG81] D. Page and C. Geilker. Indirect Evidence for Quantum Gravity. *Physical Review Letters*, **47**:979–982, 1981.

[Pit85] I. Pitowsky. On the Status of Statistical Inferences. *Synthese*, **63**(2):233–247, 1985.

[Pit89] I. Pitowsky. *Quantum Probability – Quantum Logic*. Springer Verlag, Berlin, 1989.

[Pit90] I. Pitowsky. The Physical Church Thesis and Physical Computational Complexity. *Iyyun*, **39**(1):81–99, 1990.

[Pit94] I. Pitowsky. George Boole's 'Conditions of Possible Experience' and the Quantum Puzzle. *British Journal for the Philosophy of Science*, **45**(1):95–125, 1994.

[Pit96] I. Pitowsky. Laplace's Demon Consults and Oracle: The Computational Complexity of Prediction. *Studies in the History and Philosophy of Modern Physics*, **27**(2):161–180, 1996.

[Pit07] I. Pitowsky. From Logic to Physics: How the Meaning of Computation Changed over Time. *Lecture Notes in Computer Science*, **4497**:621–631, 2007.

[Pla99] M. Planck. Über Irreversible Strahlungsvorgänge. *Sitzungsberichte der Königlich Preussischen Akademie der Wissenschaften zu Berlin*, **5**:440–480, 1899.

[Poi98] H. Poincaré. On the Foundations of Geometry. *The Monist*, **9**, 1898.

[Poi02] H. Poincaré. *Science and Hypothesis*. Flammarion, Paris, 1902. Translated by G. Halsted.

[Poi06] H. Poincaré. Sur la Dynamique de l'Electron. *Rendiconti del Circolo Matematico di Palermo*, **21**:129–175, 1906.

[Pol98] J. Polchinski. *String Theory*. Cambridge University Press, Cambridge, 1998.

[Pol07] J. Polchinski. All Strung Out. *American Scientist*, **4**:1–6, 2007.

[PR60] A. Peres and N. Rosen. Quantum Limitations on the Measurement of Gravitational Fields. *Physical Review*, **118**:335–336, 1960.

[Pre92] J. Preskill. Do Black Holes Destroy Information? *Preprint*, 1992. arXiv:hep-th/9209058v1.

[PS00] N. V. Prokof'ev and P. Stamp. Theory of the Spin Bath. *Reports on Progress in Physics*, **63**:669–726, 2000.

[PU50] A. Pais and G. Uhlenbeck. On Field Theories with Non-Localized Action. *Physical Review*, **79**:145–165, 1950.

[Put75] H. Putnam. Philosophy and Our Mental Life. In *Mind, Language, and Reality*, Vol. 2, pages 291–303. Cambridge University Press, Cambridge, 1975.

[PV49] W. Pauli and F. Villars. On the Invariant Regularization in Relativistic Quantum Theory. *Reviews of Modern Physics*, **21**:434–444, 1949.

[PVA+12] I. Pikovski, M. Vanner, M. Aspelmeyer, M. Kim, and C. Brukner. Probing Planck-Scale Physics with Quantum Optics. *Nature Physics*, **8**:393–397, 2012.

[PWS09] C. Prescod-Weinstein and L. Smolin. Disordered Locality as an Explanation for the Dark Energy. *Physical Review D*, **80**(6):063505, 2009.

[Red88] M. Redhead. A Philosopher Looks at Quantum Field Theory. In H. Brown and R. Harre, editors, *Philosophical Foundations of Quantum Field Theory*, pages 9–23. Clarendon Press, Oxford, 1988.

[Rei42] H. Reichenbach. *Philosophical Foundations of Quantum Mechanics*. Dover, New York, 1942.

[Rie73] B. Riemann. On the Hypotheses which Lie at the Bases of Geometry. *Nature*, **8**(183):14–17, 1854/1873. Translated by W. Clifford.

[Rog68] B. Rogers. On Discrete Spaces. *American Philosophical Quarterly*, **5**(2):117–123, 1968.

[Roh60] F. Rohrlich. Self-Energy and Stability of the Classical Electron. *American Journal of Physics*, **28**:639–644, 1960.

[Roh07] F. Rohrlich. *Classical Charged Particles*. World Scientific, Singapore, 2007.

[Ros30] L. Rosenfeld. Über die Gravitationswirkungen des Lichtes. *Zeitschrift für Physik*, **65**:589–599, 1930.

[Ros47] N. Rosen. Statistical Geometry and Fundamental Particles. *Physical Review*, **72**:298–304, 1947.

[Ros55] L. Rosenfeld. On Quantum Electrodynamics. In W. Pauli, editor, *Niels Bohr and the Development of Physics*, pages 70–95. McGraw-Hill, New York, 1955.

[Ros62] N. Rosen. Quantum Geometry. *Annals of Physics*, **19**:165–172, 1962.

[Ros63] L. Rosenfeld. On Quantization of Fields. *Nuclear Physics*, **40**:353–356, 1963.

[Ros66] L. Rosenfeld. Quantum Theory and Gravitation. In R. Cohen and J. Stachel, editors, *The Selected Papers of Leon Rosenfeld*, pages 599–608. Reidel, Dordrecht, 1979/1966.

[Ros84] D. Ross. Alternative Models of the Real Number Line in Physics. *International Journal of Theoretical Physics*, **23**(12):1207–1219, 1984.

[Rov93] C. Rovelli. A Generally Covariant Quantum Field Theory and a Prediction on Quantum Measurements of Geometry. *Nuclear Physics B*, **405**:797–815, 1993.

[Rov96] C. Rovelli. Black Hole Entropy from Loop Quantum Gravity. *Physical Review Letters*, **77**:3288–3291, 1996.

[Rov04] C. Rovelli. *Quantum Gravity*. Cambridge University Press, Cambridge, 2004.

[Rov07] C. Rovelli. Comment on 'Are the spectra of geometrical operators in Loop Quantum Gravity really discrete?' by B. Dittrich and T. Thiemann. *Preprint*, 2007. arXiv:gr-qc/0708.2481.

[Rov08a] C. Rovelli. A Note on DSR. *Preprint*, 2008. arXiv:gr-qc/0808.3505v2.

[Rov08b] C. Rovelli. Loop Quantum Gravity. *Living Reviews in Relativity*, **11**, 2008.

[Rov11] C. Rovelli. A Critical Look at Strings. *Foundations of Physics*, **9**:147, 2011.

[RS69] D. Reisler and N. Smith. Geometry over a Finite Field. 1969. Defense Technical Information Center OAI-PMH Repository.

[RS95] C. Rovelli and L. Smolin. Discreteness of Area and Volume in Quantum Gravity. *Nuclear Physics B*, **442**:593–619, 1995.

[RS03] C. Rovelli and S. Speziale. Reconcile Planck-Scale Discreteness and the Lorentz–FitzGerald Contraction. *Physical Review D*, **67**:064019, 2003.

[Rua28] A. Ruark. The Limits of Accuracy in Physical Measurements. *Proceedings of the National Academy of Sciences*, **14**(4):322–328, 1928.

[Rua31] A. Ruark. The Roles of Discrete and Continuous Theories in Physics. *Physical Review*, **37**:315–326, 1931.

[Rue92] A. Rueger. Attitudes Towards Infinites: Responses to Anomalies in Quantum Electrodynamics, 1927–1947. *Historical Studies in the Physical and Biological Sciences*, **22**(2):309–337, 1992.

[Rus27] B. Russell. *The Analysis of Matter*. Kegan Paul, London, 1927.

[Rus29] B. Russell. *Our Knowledge of the External World*. W.W. Norton, New York, 1929.

[Rus38] B. Russell. *The Principles of Mathematics*, second edition. W.W. Norton, New York, US, 1903/1938.

[RZ02] S. Rugh and H. Zinkernagel. The Quantum Vacuum and the Cosmological Constant Problem. *Studies in the History and Philosophy of Modern Physics*, **33**(4):663–705, 2002.

[S⁺01] C. Simon *et al.* No-Signaling Condition and Quantum Dynamics. *Physical Review Letters*, **87**:170405, 2001.

[Sal80] W. Salmon. *Space, Time and Motion*. University of Minnesota Press, Minneapolis, MN, 1980.

[Sal01] W. Salmon. *Zeno's Paradoxes*, second edition. Hacket, Indianapolis, IN, 2001.

[San05] G. Santayana. *The Life of Reason, or, The Phases of Human Progress. Introduction and Reason in Common Sense*. Scribner's, New York, 1905.

[Sau07] T. Sauer. Einstein's Unified Field Theory Program. In M. Janssen and C. Lehner, editors, *The Cambridge Companion to Einstein*. Cambridge University Press, Cambridge, 2007.

[SB99] W. Sieg and J. Byrnes. An Abstract Model for Parallel Computations. *The Monist*, **82**:150–164, 1999.

[Sch34] E. Schrödinger. Über die Unanwendbarkeit der Geometrie im Kleinen. *Die Naturwissenschaften*, **22**:518–520, 1934.

[Sch36] M. Schönberg. Letter to Mestre from 11/6/1936. In A. I. Hamburger, editor, *Obra Scientifica de Mario Schönberg; Volume 1:1936–1948*, pages xxiii–xxiv. EdUSP, Sao Paulo, 2009/1936.

[Sch48a] A. Schild. Discrete Spacetime and Integral Lorentz Transformations. *Physical Review*, **73**:414–415, 1948.

[Sch48b] J. Schwinger. On Quantum Electrodynamics and the Magnetic Moment of the Electron. *Physical Review*, **73**:416, 1948.

[Sch49a] A. Schild. Discrete Spacetime and integral Lorentz Transformations. *Canadian Journal of Mathematics*, **1**(4):29–47, 1949.

[Sch49b] P. A. Schilpp. *Albert Einstein, Philosopher-Scientist*. Open Court, New York, 1949.

[Sch57] M. Schönberg. Quantum Kinematics and Geometry. *Il Nuovo Cimento Supplement*, **6**(1):356–380, 1957.

[Sch58] J. Schwinger. Preface. In J. Schwinger, editor, *Quantum Electrodynamics*, pages viii–xvii. Dover, New York, 1958.

[Sch94] S. Schweber. *QED and the Men Who Made It*. Princeton University Press, Princeton, NJ, 1994.

[Sch01] J. Schwinger. *Quantum Mechanics: Symbolism of Atomic Measurements*. Springer, Berlin, 2001.

[Sch02] S. Schweber. Quantum Field Theory: From QED to the Standard Model. In *The Cambridge History of Science, Volume 5: The Modern Physical and Mathematical Sciences*, pages 375–393. Cambridge University Press, Cambridge, 2002.

[Sch05] B. Schroer. An Anthology of Nonlocal QFT and QFT on Noncommutative Spacetime. *Annals of Physics*, **319**:92–122, 2005.

[Sch08] B. Schroer. String Theory, the Crisis in Particle Physics and the Ascent of Metaphoric Arguments. *International Journal of Modern Physics D*, **17**(13–14):2373–2431, 2008.

[SH09]   L. Smolin and S. Hossenfelder. Conservative Solutions to the Black Hole Information Problem. *Preprint*, 2009. arXiv:0901.3156.

[Shi00]   Y. Shi. Early Gedanken Experiments of Quantum Mechanics Revisited. *Annalen der Physik*, 9:637–648, 2000.

[Sho69]   S. Shoemaker. Time Without Change. *Journal of Philosophy*, **66**:363–381, 1969.

[Sil24]   L. Silberstein. *The Theory of Relativity*, second edition. Macmillan, London, 1924.

[Sil36]   L. Silberstein. *Discrete Spacetime. A Course of Five Lectures Delivered in the McLennan Laboratory*. University of Toronto Press, Toronto, 1936.

[Smi07]   S. Smith. Continuous Bodies, Impenetrability, and Contact Interactions: The View from the Applied Mathematics of Continuum Mechanics. *British Journal of the Philosophy of Science*, **58**(3):503–538, 2007.

[Smo01]   L. Smolin. *The Trouble with Physics*. Houghton Mifflin, New York, 2001.

[Smo04]   L. Smolin. Atoms of Space and Time. *Scientific American*, January **1**:66–77, 2004.

[Smo05]   L. Smolin. The case for Background Independence. *Preprint*, 2005. arXiv/hep-th/0507235.

[Smo10]   L. Smolin. Generic Predictions of Quantum Theories of Gravity. In D. Oriti, editor, *Approaches to Quantum Gravity*, pages 548–570. Cambridge University Press, Cambridge, 2010.

[Sny47a]   H. Snyder. Quantized Spacetime. *Physical Review*, **71**(1):38–41, 1947.

[Sny47b]   H. Snyder. The Electromagnetic Field in Quantized Space–Time. *Physical Review*, **72**:68–71, 1947.

[Sor98]   R. Sorkin. The Statistical Mechanics of Black Hole Thermodynamics. In R. Wald, editor, *Black Holes and Relativistic Stars*, pages 177–194. University of Chicago Press, Chicago, IL, 1998.

[SP03]   O. Shagrir and I. Pitowsky. Physical Hypercomputation and the Church–Turing Thesis. *Minds and Machines*, **13**:87–101, 2003.

[Sta93]   J. Stachel. The Other Einstein. *Science in Context*, **6**:275–290, 1993.

[Sta02]   J. Stachel. *Einstein from 'B' to 'Z'*. Birkhauser, Boston, MA, 2002.

[Sta12]   P. Stamp. Environmental Decoherence versus Intrinsic Decoherence. *Philosophical Transactions of the Royal Society of London A*, **370**(1975):4429–4453, 2012.

[Ste77]   H. Stein. Some Philosophical Prehistory of General Relativity. In J. Earman, C. Glymour, and J. Stachel, editors, *Foundations of Spacetime Theories*, pages 3–49. University of Minnesota Press, Minneapolis, MN, 1977.

[Ste98]   E. Steinhart. Digital Metaphysics. In T. Bynum and J. Moor, editors, *The Digital Phoenix: How Computers are Changing Philosophy*, pages 117–134. Basil Blackwell, New York, 1998.

[SU03]   R. Schützhold and W. Unruh. Large-scale Nonlocality in Doubly Special Relativity with an Energy-Dependent Speed of Light. *Journal of Experimental and Theoretical Physics Letters*, **78**:431, 2003.

[Sup93]   P. Suppes. The Transcendental Character of Determinism. *Midwest Studies in Philosophy*, **18**:242–257, 1993.

[Sup01]   P. Suppes. Finitism in Geometry. *Erkenntnis*, **54**(1):133–144, 2001.

[Sus95]   L. Suskind. The World as a Hologram. *Journal of Mathematical Physics*, **36**:6377–6396, 1995.

[SV96]   A. Strominger and C. Vafa. Microscopic Origin of the Bekenstein–Hawking Entropy. *Physics Letters B*, **379**(1–4):99–104, 1996.

[SW54] E. Stueckelberg and G. Wanders. Acausalité de l'Interaction Non-locale. *Helvetica Physica Acta*, **27**:667, 1954.

[SW58] H. Salecker and E. Wigner. Quantum Limitations of the Measurement of Space–Time Distances. *Physical Review*, **109**:571, 1958.

[Swa12a] W. F. G. Swann. A Deduction of the Effects of Uniform Translatory Motion from the Electrical Theory of Matter. *Philosophical Magazine*, **23**(133):65–86, 1912.

[Swa12b] W. F. G. Swann. The FitzGerald Lorentz Contraction. *Philosophical Magazine*, **23**(133):86–94, 1912.

[Swa30] W. F. G. Swann. Relativity and Electrodynamics. *Reviews of Modern Physics*, **2**(3):243–304, 1930.

[Swa40] W. F. G. Swann. The Significance of Scientific Theories. *Philosophy of Science*, **7**:273–287, 1940.

[Swa41a] W. F. G. Swann. Philosophy of the Physical Sciences. *Nature*, **148**:692, 1941.

[Swa41b] W. F. G. Swann. The Relation of Theory to Experiment in Physics. *Reviews of Modern Physics*, **13**:190–196, 1941.

[Swa41c] W. F. G. Swann. Relativity, the FitzGerald–Lorentz Contraction. *Reviews of Modern Physics*, **13**:197–202, 1941.

[Swa60] W. F. G. Swann. Certain Matters in Relation to the Restricted Theory of Relativity, with Special Reference to the Clocks Paradox and the Paradox of the Identical Twins. *American Journal of Physics*, **28**:55–64, 319–323, 1960.

[Tel95] P. Teller. *An Interpretive Introduction to Quantum Field Theory*. Princeton University Press, Princeton, NJ, 1995.

['tH93] G. 'tHooft. Dimensional Reduction in Quantum Gravity. *Preprint*, 1993. arXiv:gr-qc/9310026.

['tH97] G. 'tHooft. Quantum Mechanical Behavior in a Deterministic Model. *Foundations of Physics Letters*, **10**:105–111, 1997.

[Tho54] J. Thomson. Tasks and Super-Tasks. *Analysis*, **15**:1–13, 1954.

[Tof84] T. Toffoli. Cellular Automata as an Alternative to (Rather than an Approximation of) Differential Equations. *Physica D*, **10**:117–127, 1984.

[Tur37] A. Turing. On Computable Numbers, with an Application to the *Entscheidungsproblem*. *Proceedings of the London Mathematical Society, Series 2*, **42**(1):230–265, 1937.

[Uff85] J. Uffink. Verification of the Uncertainty Principle in Neutron Interferometry. *Physics Letters A*, **108**(2):59–62, 1985.

[UH85] J. Uffink and J. Hilgevoord. Uncertainty Principle and Uncertainty Relations. *Foundations of Physics*, **15**(9):925–944, 1985.

[Unr81] W. Unruh. Experimental Black-Hole Evaporation? *Physical Review Letters*, **46**:1351–1353, 1981.

[Unr84] W. Unruh. Steps Towards a Quantum Theory of Gravity. In S. Christensen, editor, *Quantum Theory of Gravity*, pages 234–242. Adam Hilger, Bristol, 1984.

[Unr95] W. Unruh. Sonic Analogue of Black Holes and the Effects of High Frequencies on Black Hole Evaporation. *Physical Review D*, **51**:2827–2838, 1995.

[Unr10] W. Unruh. Minkowski Spacetime and Quantum Mechanics. In V. Petkov, editor, *Minkowski Spacetime: A Hundred Years Later*, pages 133–150. Springer, Berlin, 2010.

[Unr12] W. Unruh. Decoherence without Dissipation. *Philosophical Transactions of the Royal Society of London A*, **370**(1975):4454–4459, 2012.

[Val11] M. B. Valente. Are Virtual Quanta Nothing but Formal Tools? *International Studies in the Philosophy of Science*, **25**(1):39–53, 2011.

[Ven92] G. Veneziano. Classical and Quantum Gravity from String Theory. In *Classical and Quantum Gravity; Proceedings of the 1st Iberian Meeting on Gravity, Evora, Portugal*, pages 134–180. World Scientific, 1992.

[VB87] J. Van Bendegem. Zeno's Paradoxes and the Tile Argument. *Philosophy of Science*, **54**(2):295–302, 1987.

[VB10] J. Van Bendegem. Finitism in Geometry. In E. Zalta, editor, *Stanford Encyclopedia for Philosophy*. 2010.

[VF98] G. Vandegrift and B. Fultz. The Mössbauer effect explained. *American Journal of Physics*, **66**(7):593–596, 1998.

[vL55] C. R. von Liechtenstern. Die Beseitigung von Wlderspruchen bei der Ableitung der UnschŁrferelation. In *Proceedings of the Second International Congress of the International Union for the Philosophy of Science (Zurich 1954)*, pages 67–70. Editions du Griffon, Neuchatel, 1955.

[vN32] J. von Neumann. *Mathematical Foundations of Quantum Theory*. Princeton University Press, Princeton, NJ, 1932.

[vN36] J. von Neumann. Quantum Mechanics of Infinite Systems. In M. Redei and M. Stolzner, editors, *John von Neumann and the Foundations of Quantum Physics*, pages 249–269. Kluwer Academic Publishers, Dordrecht, 2001/1936.

[Vol03] G. Volovik. What Can the Quantum Liquid Say on the Brane Black Hole, the Entropy of an Extremal Black Hole, and the Vacuum Energy? *Foundations of Physics*, **33**(2):349–368, 2003.

[Wag93] S. Wagon. *The Banach–Tarski Paradox*. Cambridge University Press, Cambridge, 1993.

[Wal01] R. Wald. The Thermodynamics of Black Holes. *Living Reviews in Relativity*, **4**(6), 2001.

[Wal06] D. Wallace. In Defense of Naiveté: The Conceptual Status of Lagrangian Quantum Field Theory. *Synthese*, **151**:33–80, 2006.

[Wal11] D. Wallace. Taking Particle Physics Seriously: A Critique of the Algebraic Approach to Quantum Field Theory. *Studies in the History and Philosophy of Modern Physics*, **42**(2):116–125, 2011.

[Wat30a] G. Wataghin. Über die Unbestimmtheitsrelationen der Quantentheorie. *Zeitschrift für Physik*, **65**:285–288, 1930.

[Wat30b] G. Wataghin. Über eine Genauigkeitsgrenze der Ortsmessungen. *Zeitschrift für Physik*, **66**(9–10):650–651, 1930.

[Wat34a] G. Wataghin. Bemerkung über die Selbstenergie der Elektronen. *Zeitschrift für Physik*, **88**:92–98, 1934.

[Wat34b] G. Wataghin. Über die relativitiche Quantenelectrdynamik und die ausstarhlung bei Stösen sher Energierecher Elektronen. *Zeitschrift für Physik*, **92**:547–560, 1934.

[Wat36] G. Wataghin. Sulle Forze d'Inerzia Secondo la Teoria Quantistica della Gravitazione. *La Ricerca Scientifica*, **7**(2)(5–6):341, 1936.

[Wat37a] G. Wataghin. Sopra un Sistema di Equazioni Gravitazionali del Primo Ordine. *Atti della Reale Accademia Nazionale dei Lincei. Rendiconti, Classi di Scienze Fisiche, Matematiche e Naturali*, **6**(26)(11):285—289, 1937.

[Wat37b] G. Wataghin. Sulla Teoria Quantica della Gravitazione. *La Ricerca Scientifica*, **8**(2)(5–6):1–2, 1937.

[Wat38]   G. Wataghin. Quantum Theory and Relativity. *Nature*, **142**:393–394, 1938.

[Wat43]   G. Wataghin. Thermal Equilibrium Between Elementary Particles. *Physical Review*, **63**(3–4):137, 1943.

[Wat44]   G. Wataghin. Relativity and Supplementary Indeterminacy. *Physical Review*, **65**(5–6):205–205, 1944.

[Wat72]   G. Wataghin. Remarks on Some Symmetry Problems in Cosmology and in the Theory of Particles and Fields. *Lettere al Nuovo Cimento*, **4**(3):608–610, 1972.

[Wat73]   G. Wataghin. On Cosmology and Thermodynamics. *Symposia Mathematica*, **2**:163–174, 1973.

[Wat75]   G. Wataghin, 1975. Interview with Cylon Eudóxio Silva held in Rio de Janeiro. Available at CPDOC, Fundacao Getulio Vargas.

[Wei77]   S. Weinberg. The Search for Unity: Notes for a History of Quantum Field Theory. *Daedalus*, **106**(4):17–35, 1977.

[Wei78]   S. Weinberg. Critical Phenomena for Field Theorists. In A. Zichichi, editor, *Understanding the Fundamental Constituents of Matter*, pages 1–52. Springer, Berlin, 1978.

[Wei80]   S. Weinberg. Conceptual Foundations of the Unified Theory of Weak and Electromagnetic Interactions. *Reviews of Modern Physics*, **52**:515–523, 1980.

[Wei87]   S. Weinberg. Newtonianism, Reductionism, and the Art of Congressional Testimony. *Nature*, **330**:433–437, 1987.

[Wei88]   R. Weingard. Virtual Particles and the Interpretation of Quantum Field Theory. In H. Brown and R. Harre, editors, *Philosophical Foundations of Quantum Field Theory*, pages 43–58. Clarendon Press, Oxford, 1988.

[Wen33]   G. Wentzel. Über die Eigenkräfte der Elementarteilchen. *Zeitschrift für Physik*, **86**:479–494; 635–645, 1933.

[Wes02]   J. Wess. Nonabelian Gauge Theories on Noncommutative Spaces. In P. Nath and P. Zerwas, editors, *Hamburg 2002, Supersymmetry and Unification of Fundamental Interactions*, vol. 1, pages 586–599. Deutsches Elektronensynchrotron DESY, Hamburg, 2002.

[Wey49]   H. Weyl. *Philosophy of Mathematics and the Natural Sciences*. Princeton University Press, Princeton, NJ, 1949.

[Whi19]   A. Whitehead. *An Enquiry Concerning the Principles of Natural Knowledge*. Cambridge University Press, Cambridge, 1919.

[Whi29]   A. Whitehead. *Process and Reality*. Cambridge University Press, Cambridge, 1929.

[Wil75]   K. Wilson. The Renormalization Group: Critical Phenomena and the Kondo Problem. *Reviews of Modern Physics*, **47**:773–840, 1975.

[Wil06]   C. Will. The Confrontation between General Relativity and Experiment. *Living Reviews in Relativity*, **9**(3), 2006.

[Win97]   J. Winnie. Deterministic Chaos and the Nature of Chance. In J. Earman and J. Norton, editors, *The Cosmos of Science*, pages 299–324. University of Pittsburgh Press, Pittsburgh, PA, 1997.

[Wit96]   E. Witten. Reflections on the Fate of Spacetime. *Physics Today*, **49**:24, 1996.

[Wit97]   E. Witten. Duality, Spacetime, and Quantum Mechanics. *Physics Today*, **50**(5):28–33, 1997.

[Wit98] E. Witten. Anti de Sitter Space and Holography. *Advances in Theoretical and Mathematical Physics*, **2**:253–291, 1998.

[WK74] K. Wilson and J. Kogut. The Renormalization Group and the ϵ Expansion. *Physics Reports*, **12**(2):75–199, 1974.

[Woi06] P. Woit. *Not Even Wrong*. Basic Books, New York, 2006.

[Wol85] S. Wolfram. Undecidability and Intractability in Theoretical Physics. *Physical Review Letters*, **54**:735, 1985.

[Wol86] S. Wolfram. *Theory and Applications of Cellular Automata*. World Scientific, Singapore, 1986.

[Yan47] C. Yang. On Quantized Spacetime. *Physical Review*, **72**:874, 1947.

[Yuk50] H. Yukawa. Quantum Theory of Non-Local Fields. Part I. Free Fields. *Physical Review*, **77**:219–226, 1950.

[Zee03] T. Zee. *Quantum Field Theory in a Nutshell*, second edition. Princeton University Press, Princeton, NJ, 2003.

[Zim96] D. Zimmerman. Indivisible Parts and Extended Objects: Some Philosophical episodes from Topology's Prehistory. *The Monist*, **79**(1):148–180, 1996.

[Zwi08] B. Zwiebach. *A First Course in String Theory*. Cambridge University Press, Cambridge, 2008.

Index